A Short Guide to Writing about Art

SIXTH EDITION

A Short Guide to Writing about Art

SIXTH EDITION

SYLVAN BARNET
Tufts University

 LONGMAN

An imprint of Addison Wesley Longman, Inc.

New York • Reading, Massachusetts • Menlo Park, California • Harlow, England
Don Mills, Ontario • Sydney • Mexico City • Madrid • Amsterdam

English Editor: Lynn M. Huddon
Marketing Manager: Renée Ortbals
Project Manager: Bob Ginsberg
Design Manager: Rubina Yeh
Cover Illustrations: (*Front cover*) Mary Stevenson Cassatt (U.S., 1844–1926), "In the
    Loge," 1879. Oil on canvas, 32 x 26 in. (81.3 x 66 cm.). Museum of Fine Arts,
    Boston. The Hayden Collection. 10.35. (*Back cover*) Robert Smithson, "Spiral
    Jetty," Great Salt Lake, Utah, April 1970. Photograph by Gianfranco Gorgoni.
    Copyright © Estate of Robert Smithson/Licensed by VAGA, New York, NY.
Photo Researcher: Photosearch, Inc.
Technical Desktop Manager: Heather A. Peres
Senior Print Buyer: Hugh Crawford
Electronic Page Makeup: Allentown Digital Services
Printer and Binder: RR Donnelley & Sons Company
Cover Printer: Coral Graphic Services, Inc.

Library of Congress Cataloging-in-Publication Data

Barnet, Sylvan.
    A short guide to writing about art / Sylvan Barnet. — 6th ed.
        p.   cm. — (The short guide series)
    Includes bibliographical references and index.
    ISBN 0-321-04605-6
    1. Art criticism—Authorship. I. Title.
    N7476.B37   1999
    808'.0667—dc21   99-28221
                                                      CIP

Please visit our website at http://www.awlonline.com

ISBN 0-321-04605-6

2345678910—DOC—02010099

To the memory of my brother, Howard

Contents

Preface

Another book for the student of art to read? I can only echo William James's report of the unwed mother's defense: "It's such a *little* baby."

Still, a few additional words may be useful. Everyone knows that students today do not write as well as they used to. Probably they never did, but it is a truth universally acknowledged (by English teachers) that the cure is *not* harder work from instructors in composition courses; rather, the only cure is a demand, on the part of the entire faculty, that students in all classes write decently. But instructors outside of departments of English understandably say that they lack the time—and perhaps the skill—to teach writing in addition to, say, art.

This book may offer a remedy. Students who read it—and it is short enough to be read in addition to whatever texts the instructor regularly requires—should be able to improve their essays

- by getting ideas both about works of art and about approaches to art, from the first four chapters ("Writing about Art," "Analysis," "Writing a Comparison," "How to Write an Effective Essay"), and from Chapter 6 ("Some Critical Approaches")
- by studying the principles on writing explained in Chapter 5, "Style in Writing" (e.g., on tone, paragraphing, and concreteness), and Chapters 7, 8, and 9 ("Art-Historical Research," "Writing a Research Paper," and "Manuscript Form")
- by studying the short models throughout the book, which give the student a sense of some of the ways in which people talk about art

As Robert Frost said, writing is a matter of having ideas. This book tries to help students to have ideas by suggesting questions they may ask themselves as they contemplate works of art. After all, instructors want papers that *say* something, papers with substance, not papers whose only virtue is that they are neatly typed and that the footnotes are in the proper form.

One is reminded of a story that Giambologna (1529–1608) in his old age told about himself. The young Flemish sculptor (his original name was Jean de Boulogne), having moved to Rome, went to visit the aged Michelangelo. To show what he could do, Giambologna brought with him a carefully finished, highly polished wax model of a sculpture. The master took the model, crushed it, shaped it into something very different from Giambologna's original, and handed it back, saying, "Now learn the art of modeling before you learn the art of finishing." This story about Michelangelo as a teacher is harrowing, but it is also edifying (and it is pleasant to be able to say that Giambologna reportedly told it with pleasure). The point of telling it here is not to recommend a way of teaching; the point is that a highly finished surface is all very well, but we need substance first of all. A good essay, to repeat, says something.

A *Short Guide to Writing about Art* contains notes and two sample essays by students, an essay by a professor, and numerous model paragraphs by students and by published scholars such as Rudolf Arnheim, Albert Elsen, Mary D. Garrard, Anne Hollander, and Leo Steinberg. These discussions, as well as the numerous questions that are suggested, should help students to understand the sorts of things one says, and the ways one says them, when writing about art. After all, people *do* write about art, not only in the classroom but in learned journals, catalogs, and even in newspapers and magazines.

A NOTE ON THE SIXTH EDITION

I have been in love with painting ever since I became conscious of it at the age of six. I drew some pictures which I thought fairly good when I was fifty, but really nothing I did before the age of seventy was of any value at all. At seventy-three I have at last caught every aspect of nature—birds, fish, animals, insects, trees, grasses, all. When I am eighty I shall have developed still further, and will really master the secrets of art at ninety. When I reach one hundred my art will be truly sublime, and my final goal will be attained around the age of one hundred and ten, when every line and dot I draw will be imbued with life.

—Hokusai (1760–1849)

Probably all artists share Hokusai's self-assessment. And so do all writers of textbooks. Each edition of this book seemed satisfactory to me when I sent the manuscript to the publisher, but with the passing not of decades but of only a few months I detected inadequacies, and I wanted to say new things. This sixth edition, therefore, not only includes sixth thoughts about many topics discussed in the preceding editions but it also introduces new topics.

The emphasis is still twofold—on *seeing* and *saying*, or on getting ideas about art (Chapters 1–4) and on presenting those ideas effectively in writing (Chapters 5–8)—but this edition includes new thoughts about these familiar topics, as well as thoughts about new topics. For instance, the pages concerned with generating ideas contain new material about

- the canon
- cultural materialism
- queer theory
- realism and idealism
- critical values
- the uses of the Internet

The pages concerned with effective writing contain

- boxed summaries, each with "A Rule for Writers"
- seven checklists for revising paragraphs, writing a comparison, evaluating a web site, and researching on the internet

and the discussion of documentation now includes

- *Chicago Manual* Style
- the *Art Bulletin Style Guide*
- forms used for Asian names
- citations of electronic sources

Eleven illustrations are new, including Segal's *The Diner*, Paik's *TV Buddha*, Brancusi's *Torso*, and a photograph of the Guggenheim Bilbao.

Much of the new material concerned with generating ideas responds to relatively new trends in the study of art. Today an interest in political, economic, and social implications of art has in large measure replaced the earlier interest in matters of style, authenticity, and quality. In short, contemporary interest seems to have moved from the *text* to the *context,* from the artwork as a unique object with its distinctive meaning to the artwork as a manifestation of something more important (gender, politics, ethnicity), from aesthetics to a criticism of social and political cultures. This shift in the study of art is a response to a shift in art itself—the shift from Modernism to Post-Modernism. In the first half of the twentieth century, art—in the movement called Modernism—sloughed off the earlier concern with subject matter, illusionism, and beauty; what counted was the artist's sensibility. Post-Modernism, rejecting this elite sensibility, sees artists as deeply embedded in their society, understood only in the context of that society. The emphasis is now on the historical conditions governing the production and consumption of art.

Nevertheless, *A Short Guide* continues to give generous space to the formal analysis of art. I continue to use the term *art* rather than *visual culture*, though I uneasily recall Andy Warhol's observation that in America most people think that Art is a man's name. I grant, too, that *visual culture* has the advantage of including works—for instance, boomerangs, nose rings, and Native American feathered bonnets—that we might call *art* but that are not called art by the cultures that produced them. Indeed, one has only to do a very little reading to learn that many languages do not include a word for *art;* apparently no Native American language has such a word, and the Japanese invented such a word only after coming into contact with European ideas. My use of *art*, then, should be considered not only affection for an old word but also shorthand for *visual culture*.

ACKNOWLEDGMENTS

I am fortunate in my many debts. James Cahill, Sarah Blick, Madeline Harrison Caviness, Robert Herbert, Naomi Miller, and Elizabeth de Sabato Swinton generously showed me some of their examinations, topics for essays, and guidelines for writing papers. I have received invaluable help also from those who read part or all of the manuscript of the first edition, or who made suggestions while I was preparing the revised editions. The following people called my attention to omissions, excesses, infelicities, and obscurities: Jane Aaron, Mary Clare Altenhofen, Howard Barnet, Peter Barnet, Mark H. Beers, Pat Bellanca, Morton Berman, Sarah Blick, William Burto, Ruth Butler, James Cahill, William E. Cain, Charles Christensen, Fumiko Cranston, Whitney Davis, Margaret Fields Denton, Eugene Dwyer, Elizabeth ten Grotenhuis, Maxwell Hearn, Julius Held, Joseph M. Hutchinson, Eugene J. Johnson, Deborah Martin Kao, Laura Kaufman, Leila Kinney, Susan Kuretsky, Elizabeth Anne McCauley, Sara J. MacDonald, Janice Mann, Jody Maxmin, Lawrence Nees, John O'Brian, Elizabeth Pilliod, Kenneth J. Procter, Jennifer Purtle, Patricia Rogers, John M. Rosenfield, James M. Saslow, John M. Schnorrenberg, Jack J. Spector, Marcia Stubbs, Ruth Thomas, Gary Tinterow, Stephen K. Urice, Jonathan Weinberg, Tim Whalen, and Paul J. Zelanski. I have adopted many of their statements verbatim.

The extremely generous contributions of Leila W. Kinney, Anne Mc-Cauley, and James M. Saslow must be further specified. Kinney wrote the material on the Internet, McCauley the material on photography, and Saslow the material on gay and lesbian art criticism. In each instance the job turned out to be more time-consuming than they or I had anticipated, and I am deeply grateful to them for staying with it. I am also indebted to the College Art Association, which granted me permission to reprint the *Art Bulletin Style Guide,* and to Marcia Stubbs, for letting me use some material that had appeared in a book we collaborated on, Barnet and Stubbs's *Practical Guide to Writing.*

I wish also to thank the reviewers, whose comments helped me to revise this edition: Rebecca Butterfield, University of Pennsylvania; Richard Carp, Northern Illinois University; Janet Carpenter, San Jose State University; Gail Geiger, University of Wisconsin-Madison; Dianne Goode, University of Texas at Dallas; Anna Hammond, Museum of Modern Art; Samantha Kavky, University of Pennsylvania; Arturo Lindsay, Spelman College; Robert Munman, University of Illinois at Chicago; Sheryl Reiss, Cornell University; Leland Roth, University of Oregon; Annie Shaver-Crandell, City College of New York; and Connie Stewart, University of Northern Colorado.

Finally, Bob Ginsberg, and Lynn Huddon at Addison Wesley Longman were always receptive, always encouraging, and always helpful.

SYLVAN BARNET

I saw the things which have been brought to the King from the new golden land: a sun all of gold a whole fathom broad, and a moon all of silver of the same size, also two rooms full of the armour of the people there, and all manner of wondrous weapons of theirs, harness and darts, wonderful shields, strange clothing, bedspreads, and all kinds of wonderful objects of various uses, much more beautiful to behold than prodigies. These things were all so precious that they have been valued at one hundred thousand gold florins. All the days of my life I have seen nothing that has gladdened my heart so much as these things, for I saw amongst them wonderful works of art, and I marvelled at the subtle talents of men in foreign lands. Indeed, I cannot express all that I thought there.

—Albrecht Dürer, in a journal entry of 27 August 1520,
writing about Aztec treasures sent by Motecuhzoma
to Cortés in 1519, and forwarded
by Cortés to Charles V

Painting cannot equal nature for the marvels of mountains and water, but nature cannot equal painting for the marvels of brush and ink.

—Tung-Ch'i-chang (1555–1636)

What you see is what you see.

—Frank Stella, in an interview, 1966

1

Writing about Art

Writing about *what*? What is art? The first paragraph of a book on relatively recent art includes these sentences:

> Ordinary viewers of today, hoping for coherence and beauty in their imaginative experiences, confront instead works of art declared to exist in arrangements of bare texts and unremarkable photographs, in industrial fabrications revealing no evidence of the artist's hand, in mundane commercial products merely transferred from shopping mall to gallery, or in ephemeral and confrontational performances in which mainstream moral values are deliberately travestied.
>
> —Thomas Crow, *The Rise of the Sixties: American and European Art in the Era of Dissent 1955–69* (1996), 7

Again, what is art? Perhaps we can say that art is anything that is said to be art by people who ought to know. Who are these people? They are the men and women who teach in art and art history departments, who write about art for newspapers and magazines and scholarly journals, who think of themselves as art collectors, who call themselves art dealers, and who run museums. One of the most ardently discussed items at the Whitney Museum's 1997 Biennial was David Hammon's video of a can being kicked down the Bowery. At the Dia Center for the Arts in Chelsea, Tracey Moffatt's video of surfers in a parking lot changing into swimwear, shielded by towels, created excitement. At the New Museum, Mona Hatoum's videos of the inside of her body—she sends a microvideo through one bodily orifice or another—still get lots of attention. The people who run art museums show these videos, and the people who visit the museums enjoy them, so presumably the videos are art.

Of course museum curators, museum-goers, art teachers, and all the rest change their ideas over time. For instance, until a decade or two ago, such Native American objects as blankets, headdresses, beaded clothes, and horn spoons were regarded as artifacts, not art, and consequently they were found not in art museums but in ethnographic museums. Today curators of art museums are eager to acquire and display such Native American objects. Similarly, although sculptures from sub-Saharan

1

Africa have been found in art museums since the early twentieth century, other African works—for instance, textiles, pottery, baskets, and jewelry—did not move from ethnographic museums to art museums until about 1970. Even today, however, the African objects most sought by museums are ones that show no foreign influence, except for older works, such as ivory carved for export to Portugal in the sixteenth century. Objects showing European influence or objects made for the tourist trade are rarely considered art by those who run art museums. The museums (and the museum-goers) of tomorrow may have a different idea about such objects. After all, the sixteenth-century ivories carved for the Portuguese were essentially tourist art. Maybe only our present cultural prejudice keeps most museum curators from regarding airport art as worth serious consideration.

Do snapshots—not the work of esteemed photographers but the sorts of images you have in an album or a shoe box—deserve serious consideration as art? In 1998 the San Francisco Museum of Modern Art displayed snapshots taken by anonymous amateurs in an exhibition entitled "Snapshots: The Photography of Everyday Life." The museum called it "the first exhibition to explore amateur domestic photography in a museum context, examining the borderline between what museums display and visitors might themselves create with a camera." Weston Naef, chief curator of photography at the Getty Museum in Los Angeles, and himself a collector of snapshots, isn't sure whether the snapshots are art or not:

> Hardly any of these pictures equal the depth of understanding and clarifying of formidable visual problems that the best works in the Getty's collection show. In my mind they were artifacts until the museum decided to exhibit them. Now somebody else has to prove they're *not* art.
> —Quoted in Tessa DeCarlo, "Look at the Camera, Smile . . . Got It! A Work of Art!" *The New York Times,* 5 July 1998, sec. 2, 30

And in listening to people who talk about art, let's not forget the opinions of the people who consider themselves artists. If someone with an established reputation as a painter says of a postcard she has just written, "This is a work of art," well, we probably have to be very careful before we reply, "No, it isn't." A common definition today is, "Art is what artists do," and they do a great many things that do not at all resemble Impressionist paintings. But artists too may be uncertain about what is art. An exhibition catalog, *Jackson Pollock: Black and White* (1969), reports an interesting episode. Pollock's wife, Lee Krasner, a painter herself, is quoted as saying, "In front of a very good painting . . . he asked me, 'Is this a painting?' Not is this a good painting, or a bad one, but a *painting!* The degree of doubt was unbelievable at times" (page 8).

Pollock was a highly innovative painter (for a photograph of him at work, see page 269), but sculptors too have produced highly innovative work, work that may seem not to qualify as art. Take, for instance, *earth-works* or *Earth Art,* large sculptural forms made of earth and rocks. An example is Robert Smithson's *Spiral Jetty,* created in 1970 (below and on the back cover). Smithson supervised the construction of a jetty—if a spiral can be regarded as a jetty—some 15 feet wide and 1,500 feet long, in Great Salt Lake, Utah (see below). Because the water level has since risen, *Spiral Jetty* is now submerged, though the work still survives—under water, in a film Smithson made during the construction of the jetty, and in photographs taken before the water level rose. Is this combination of mud, rocks, and water art? Smithson said so, and the writers of books on recent art agree, since they include photographs of *Spiral Jetty.* When you think about it, Earth Art is not unprecedented; photographs of the Egyptian pyramids have for decades appeared in books of art.

Let's look briefly at a work produced in 1972 by a student in the Feminist Art Program at the California Institute of the Arts and exhibited

Robert Smithson, *Spiral Jetty,* 1970. (Photograph by Gianfranco Gorgoni/Contact, Courtesy John Weber Gallery, New York © Estate of Robert Smithson/Licensed by VAGA, New York, NY.)

again at the Bronx Museum of the Arts in 1995. Two instructors and some twenty students in the class decided to take an abandoned house and turn it into a work of art, *Womanhouse*. Each participant took some part of the house—a room, a hallway, a closet—and transformed it in accord with her dreams and fantasies. The students were encouraged to make use of materials considered trivial and associated with women, such as dolls, cosmetics, sanitary napkins, and crocheted material. One student, Faith Wilding, constructed a rope web to which crochet was attached, thereby creating what she called (in 1972) *Web Room* or *Crocheted Environment* and (in the 1995 version) *Womb Room* (see below). Traditionally, a work of art (say, a picture hanging on the wall or a statue standing on a pedestal) is set apart from the spectator and is an object to be looked at from a relatively detached point of view. But *Womb Room* is a different sort of thing. It is an *installation*—a work that takes over or transforms a space, indoors or outdoors, and that usually gives the viewer a sense of being not only a spectator but also a participant in the work.

Faith Wilding crocheting the *Womb Room* installation (1995) at the Bronx Museum of the Arts. (© C. M. Hardt/NYT Pictures)

With its nontraditional material—who ever heard of making a work of art out of rope and pieces of crochet?—its unusual form, and its suggestions of the womb, a nest, and primitive architecture, Wilding's installation would hardly have been regarded as art before, say, the mid-twentieth century.

We have been talking about the idea that something is a work of art if its creator—whether a person or a culture—says it is art. But some cultures do not want some of their objects to be thought of as art. For example, although curators of American art museums have exhibited Zuni war god figures (or *Ahayu:da*), the Zuni consider such figures to be embodiments of sacred forces, not aesthetic objects, and therefore *un*suitable for exhibition. The proper place for these figures, the Zuni say, is in open-air hillside shrines.° (A question: Can we call something *art* if its creator did not think of it as art?)

What sorts of things you will write about will depend partly on your instructor, partly on the assignment, partly on what the museums in your area call art, and partly on what you call art.

WHY WRITE ABOUT ART?

We write about art in order to clarify and to account for our responses to works that interest or excite or frustrate us. In putting words on paper we have to take a second and a third look at what is in front of us and at what is within us. Picasso said, "To know what you want to draw, you have to begin drawing"; similarly, writing is a way of finding what you want to write, a way of learning.

°See Steven Talbot, "Desecration and American Indian Religious Freedom," *Journal of Ethnic Studies* 12:4 (1985): 1–8; T. J. Ferguson and B. Martza, "The Repatriation of Zuni *Ahayu:da*," *Museum Anthropology* 14:2 (1990): 7–15. For additional discussions of the social, political, and ethical questions that face curators, see Moira Simpson, *Making Representations: Museums in the Post-Colonial Era* (1996); *Exhibiting Dilemmas: Issues of Representation at the Smithsonian*, ed. Amy Henderson and Adrienne L. Kaeppler (1997). Some authors of books go so far as not to reproduce certain images in deference to the wishes of the community. Example: Janet C. Berlo and Ruth B. Phillips, in *Native North American Art* (1998), inform readers that a certain kind of Iroquois mask, representing forest spirits, is not illustrated because these masks "are intended only to be seen by knowledgeable people able to control these powers" (page 11). The heart of the issue perhaps may be put thus: Is it appropriate for one culture to take the sacred materials of another culture out of their context and to exhibit them as aesthetic objects to be enjoyed?

The last word is never said about complex thoughts and feelings—and works of art, as well as our responses to them, embody complex thoughts and feelings. But when we write about art we hope to make at least a little progress in the difficult but rewarding job of talking about our responses. As Arthur C. Danto says in the introduction to *Embodied Meanings* (1994), a collection of essays about art:

> Until one tries to write about it, the work of art remains a sort of aesthetic blur. . . . After seeing the work, write about it. You cannot be satisfied for very long in simply putting down what you felt. You have to go further. (14)

When we write, we learn; we also hope to interest our reader by communicating our responses to material that for one reason or another is worth talking about.

But to respond sensitively to anything and then to communicate responses, we must have some understanding of the thing, and we must have some skill at converting responses into words. This book tries to help you deepen your understanding of art—what art does and the ways in which it does it—and the book also tries to help you transform your responses into words that will let your reader share your perceptions, your enthusiasms, and even your doubts. This sharing is, in effect, teaching. Students often think that they are writing for the instructor, but this is a misconception; when you write, *you* are the teacher. An essay on art is an attempt to help someone to see the work as you see it.

THE WRITER'S AUDIENCE AS A COLLABORATOR

If you are not writing for the instructor, for whom are you writing? For yourself, of course, but also for an audience that you must imagine. All writers need to imagine some sort of audience: Writers of self-help books keep novices in mind, writers of articles for *Time* keep the general public in mind, writers of papers for legal journals keep lawyers in mind, and writers of papers for *The Art Bulletin* keep art historians in mind. An imagined audience in some degree determines what the writer will say—for instance, it determines the degree of technical language that may be used and the amount of background material that must be given.

Who is *your* audience? In general (unless your instructor suggests otherwise) think of your audience as your classmates. If you keep your

✍ A RULE FOR WRITERS:

When you draft and revise, keep your audience in mind.

classmates in mind as your audience, you will not write, "Leonardo da Vinci, a famous Italian painter," because such a remark offensively implies that the reader does not know Leonardo's nationality or trade. You might, however, write, "Leonardo da Vinci, a Florentine by birth," because it's your hunch that your classmates do *not* know that Leonardo was born in Florence, as opposed to Rome or Venice. And you *will* write, "John Butler Yeats, the expatriate Irish painter who lived in New York," because you are pretty sure that only specialists know about Yeats. Similarly, you will *not* explain that the Virgin Mary was the mother of Jesus, but you probably will explain that St. Anne was the mother of Mary.

In short, if you imagine that your reader is looking over your shoulder when you are revising, your imagined audience becomes your collaborator, helping you to decide what you need to say—in particular helping you to decide

- which terms you need to define
- what degree of detail you need to go into

If, for instance, you are offering a psychoanalytic interpretation, you can assume that your audience is familiar with the name Freud and with the Oedipus Complex, but you probably cannot assume that your audience is familiar with the contemporary psychoanalyst D. W. Winnicott and his concept of the pre-Oedipal mother-infant dyad as a source of creativity. If you are going to make use of Winnicott, you will have to identify him and briefly explain his ideas.

A successful essay, whether a one-page review of an art exhibition in *Time* or a twenty-page essay in *The Art Bulletin,* begins with where the readers are, and then goes on to take the readers further.

THE FUNCTIONS OF CRITICAL WRITING

In everyday language the most common meaning of criticism is "finding fault," and to be critical is to be censorious. But a critic can see excellences as well as faults. Because we turn to criticism with the hope that the critic has seen something we have missed, the most valuable criticism is not that which shakes its finger at faults but that which calls our attention to interesting matters going on in the work of art. In the following statement W. H. Auden suggests that criticism is most useful when it calls our attention to things worth attending to. He is talking about works of literature, but we can easily adapt his words to the visual arts.

> What is the function of a critic? So far as I am concerned, he can do me one or more of the following services:
>
> 1. Introduce me to authors or works of which I was hitherto unaware.
> 2. Convince me that I have undervalued an author or a work because I had not read them carefully enough.
> 3. Show me relations between works of different ages and cultures which I could never have seen for myself because I do not know enough and never shall.
> 4. Give a "reading" of a work which increases my understanding of it.
> 5. Throw light upon the process of artistic "Making."
> 6. Throw light upon the relation of art to life, to science, economics. ethics, religion, etc.
>
> —W. H. Auden, *The Dyer's Hand* (1963), 8–9

The emphasis on observing, showing, illuminating suggests that the chief function of critical writing is not very different from the common view of the function of literature or art. The novelist Joseph Conrad said that his aim was "before all, to make you *see*," and the painter Ben Shahn said that in his paintings he wanted to get right the difference between the way a cheap coat and an expensive coat hung.

Take Auden's second point, that a good critic can convince us—can gain our agreement by calling attention to evidence supporting a thesis—that we have undervalued a work. Although you probably can draw on your own experience for confirmation, an example may be useful. Rembrandt's self-portrait with his wife (see page 9), now in Dresden, strikes many viewers as one of his least attractive pictures: the gaiety seems forced, the presentation a bit coarse and silly. Paul Zucker, for example, in *Styles in Painting*, finds it "over-hearty," and John Berger, in *Ways of Seeing*, says that "the painting as a whole remains an advertisement for the sitter's good fortune, prestige, and wealth. (In this case Rembrandt's own.) And like all such advertisements it is heartless." But some scholars have pointed out, first, that this picture may be a representation of the Prodigal Son, in Jesus's parable, behaving riotously, and, second, that it

Rembrandt, *Self-Portrait with Saskia*, ca. 1635. Oil on canvas, 5′4″ × 4′4″. (Art Resource, NY/Erich Lessing; Gemäldegalerie, Staatliche Kunstsammlungen, Dresden, Germany/Art Resource, NY)

may be a profound representation of one aspect of Rembrandt's marriage. Here is Kenneth Clark on the subject:

> The part of jolly toper was not in his nature, and I agree with the theory that this is not intended as a portrait group at all, but as a representation of the Prodigal Son wasting his inheritance. A tally-board, faintly discernible on the left, shows that the scene is taking place in an inn. Nowhere else has Rembrandt made himself look so deboshed, and Saskia is enduring her ordeal with complete detachment—even a certain hauteur. But beyond the ostensible subject, the picture may express some psychological need in Rembrandt to reveal his discovery that he and his wife were two very different characters, and if she was going to insist on her higher social status, he would discover within himself a certain convivial coarseness.
>
> —Kenneth Clark, *An Introduction to Rembrandt* (1978), 73

After reading these words we may find that the appeal of the picture grows—and any analysis that increases our enjoyment in a work surely serves a useful purpose. Clark's argument, of course, is not airtight—one rarely can present an airtight argument when writing about art—but notice that Clark does more than merely express an opinion or report a feeling. In his effort to persuade us, he offers evidence (the tally-board, and the observation that no other picture shows Rembrandt so "deboshed"), and the evidence is strong enough to make us take another look at the picture. After looking again, we may come to feel that we have undervalued the picture.

A SAMPLE ESSAY

Kenneth Clark's paragraph comes from one of his two books on Rembrandt. Clark's audience is not limited to art historians, but it is limited to the sort of person who might read a book about Rembrandt. The following essay on Jean-François Millet's *The Gleaners,* written by Robert Her-

Jean-François Millet, *The Gleaners,* 1857. Oil on canvas, 32 ⅞" × 43¼". (Service Photographique de la Réunion des Musées Nationaux/The Musée d'Orsay, Paris)

bert, was originally a note in the catalog issued in conjunction with the art exposition at the Canadian World's Fair, Expo 67. Herbert's audience thus is somewhat wider and more general than Clark's. Given his audience, Herbert reasonably offers not a detailed study of one aspect of the painting, say, its composition; rather, he performs most of the services that on page 8 Auden says a critic can perform. In this brief essay, Herbert skillfully sets forth material that might have made half a dozen essays: Millet's life, the background of Millet's thought, Millet's political and social views, the composition of *The Gleaners,* Millet's depiction of peasants, Millet's connection with later painters. But the aim is always to make us *see.* In *The Gleaners* Millet tried to show us certain things, and now Robert Herbert tries to show us—tries to make us see—what Millet was doing and how he did it.

Robert Herbert
MILLET'S *THE GLEANERS*

Jean-François Millet, born of well-to-do Norman peasants, began his artistic training in Cherbourg. In 1837 he moved to Paris where he lived until 1849, except for a few extended visits to Normandy. With the sounds of the Revolution of 1848 still rumbling, he moved to Barbizon on the edge of the Forest of Fontainebleau, already noted as a resort of landscape painters, and there he spent the rest of his life. One of the major painters of what came to be called the Barbizon School, Millet began to celebrate the labors of the peasant, granting him a heroic dignity which expressed the aspirations of 1848. Millet's identification with the new social ideals was a result not of overtly radical views, but of his instinctive humanitarianism and his rediscovery in actual peasant life of the eternal rural world of the Bible and of Virgil, his favorite reading since youth. By elevating to a new prominence the life of the common people, the revolutionary era released the stimulus which enabled him to continue this essential pursuit of his art and of his life.

The Gleaners, exhibited in the Salon of 1857, presents the very poorest of the peasants who are fated to bend their backs to gather with clubbed fingers the wisps of overlooked grain. That they seem so entirely wedded to the soil results from the perfect harmony of Millet's fatalistic view of man with the images which he created by a careful disposition of lines, colors, and shapes.

The three women are alone in the bronzed stubble of the foreground, far removed from the bustling activity of the harvesters in the distance, the riches of whose labors have left behind a few gleanings. Millet has weighted his figures ponderously downward, the busy harvest scene is literally above them, and the high horizon line which the taller woman's cap just touches emphasizes their earth-bound role, suggesting that the sky is a barrier which presses down upon them, and not a source of release.

The humility of primeval labor is shown, too, in the creation of primitive archetypes rather than of individuals. Introspection such as that seen in Velázquez's *Water Carrier of Seville,* in which the three men are distinct individuals, is denied by suppressing the gleaners' features, and where the precise, fingered gestures of La Tour's *Saint Jerome* bring his intellectual work toward his sensate mind, Millet gives his women clublike hands which reach away from their bent bodies toward the earth.

It was, paradoxically, the urban-industrial revolution in the nineteenth century which prompted a return to images of the preindustrial, ageless labors of man. For all their differences, both Degas and Van Gogh were to share these concerns later, and even Gauguin was to find in the fishermen of the South Seas that humble being, untainted by the modern city, who is given such memorable form in Millet's *Gleaners.*

The Thesis and the Organization

This essay includes **evaluation,** or judgment, as well as analysis of what is going on in the painting. First, Robert Herbert judges Millet's picture as worth talking about. (Newspaper and magazine criticism is largely concerned with evaluation—think, for instance, of the film review, which exists chiefly to tell the viewer whether a film is worth seeing—but most academic criticism assumes the value of the works it discusses, and it is chiefly analytic and interpretive. For a more detailed comment on evaluation, see pages 174–85.) Although Herbert explicitly praises some of the work's qualities (its "perfect harmony" and "memorable form"), most of his evaluation is implicit in and subordinate to the analysis of what he sees. (For the moment we can define **analysis** as the separation of the whole into its parts; the second chapter of this book is devoted to the topic.) The essayist sees things and calls them to our attention as worthy of note. He points out the earthbound nature of the women, the difference between their hands and those of Saint Jerome (in another picture that was in the exhibition), the influences of the Bible and of Virgil, and so forth. It is clear that he values the picture, and he states some of the reasons he values it; but he is not worried about whether Millet is a better artist than Velázquez, or whether this is Millet's best painting. He is content to help us see what is going on in the picture.

Or at least he seems to be content to help us see. In fact, Herbert is advancing a **thesis** (a central point, a main idea)—in this case that the picture celebrates the heroic dignity of the peasant. A good thesis is *not* the assertion of an undisputed fact (*"The Gleaners* was painted in 1857"), nor is it a broad generalization that cannot interestingly be supported (*"The Gleaners* is widely admired"). Normally, a thesis statement *names the topic* (here, a particular painting) *and makes an assertion about it that the writer will support with details later in the essay.* Robert Herbert's **thesis statement,** appropriately set forth in his first paragraph, is

> Millet began to celebrate the labors of the peasant, granting him a heroic dignity which expressed the aspirations of 1848.

In this case, Herbert does not name the topic, but he has already done so in his title. In the rest of his essay he tries to *persuade* us, by offering an argument—a reasoned account, consisting of evidence offered in support of the thesis—that his thesis is valid. He has seen something, and he wants us to see it too. Notice that he sees with more than his eyes: Memories, emotions, knowledge, and value systems help him to see, and his skill as a writer helps him to persuade us of the accuracy of his report. If he wants to convince us and to hold our interest, he has to do more than offer random perceptions; he has to present his perceptions coherently.

It is not enough for writers to see things and to report to readers what they have seen. Writers have to present their material in an orderly fashion, so that readers can take it in and can follow a developing argument. In short, writers must organize their material. Let's look for a moment at the **organization,** or plan, of Herbert's essay. In his effort to help us see what is going on, the author keeps his eye on his subject.

- The opening paragraph includes a few details (e.g., the fact that Millet was trained in Cherbourg) that are not strictly relevant to his main point (the vision embodied in the picture) but that must be included because the essay is not only a critical analysis of the picture but an informative headnote in a catalog of an exhibition of works by more than a hundred artists. Even in this preliminary biographical paragraph the writer moves quickly to the details closely related to the main business: Millet's peasant origin, his early association with landscape painters, his humanitarianism, and his reading of the Bible and Virgil.

- The second paragraph takes a close look at some aspects of the picture (the women's hands, their position in the foreground, the harvesters above and behind them, the oppressive sky). Notice too that this paragraph calls attention to the social context: Herbert points out that the "poorest of the peasants" are apart "from the bustling activity of the harvesters in the distance."

- The third paragraph makes illuminating comparisons with two other paintings in the exhibition. (A good description—one that catches the individuality of a particular work—almost always makes use of comparison. On comparing, see pages 101–15.)

- The last paragraph, like most good concluding paragraphs, while recapitulating the main point (the depiction of ageless labors), enlarges the vision by including references to Millet's younger contemporaries who shared his vision. Notice that this new material does not leave us looking forward to another paragraph but neatly opens up, or enriches, the matter and then turns it back to Millet. (For additional remarks on introductory and concluding paragraphs, see pages 142–47.)

A Note on Outlining

If Herbert prepared an outline to help him draft his essay, it may have looked something like this:

Biographical background
   Barbizon
   Millet's humanitarianism
<u>The Gleaners</u>
   The figures and the setting (poorest peasants apart from harvesters)
   Compare (contrast) with Velázquez and La Tour
Concluding paragraph
   Millet looks back, but anticipates later painters (Degas, van Gogh)

An outline—nothing elaborate, even just a few notations in a sequence that seems reasonable—can be a great help in drafting an essay. The very act of putting a few ideas down on paper will usually stimulate you to think of additional ideas, just as when you jot down "tuna fish" on a shopping list, you are somehow reminded that you also need to pick up bread.

✍ **A RULE FOR WRITERS:**

Organize your essay so that your readers can easily follow the argument you use to support your thesis.

Outlining, in short, is not merely a way of organizing ideas but is also a way of getting ideas.

If, however, you feel that you can't make a preliminary outline for a draft, write a draft and *then* outline it. Why? An outline of your draft will let you easily examine your organization. That is, when in the course of reviewing your draft you brush aside the details and put down the chief point or basic idea of each paragraph, you will produce an outline that will help you to see if you have set forth your ideas in a reasonable sequence, a sequence that will help rather than confuse your readers. By studying this outline of your draft you may find, for instance, that your third point would be better used as your first point. (Outlining is discussed in more detail on pages 92–93 and 119–21.)

WHAT IS AN INTERPRETATION—AND ARE ALL INTERPRETATIONS EQUALLY VALID?

Interpretation and Interpretations

We can define **interpretation** as a setting forth of the meaning of a work of art, or, better, as the setting forth of one of the meanings of the work. This issue of *meaning* versus *meanings* deserves a brief explanation. Although some art historians still believe that a work of art has a single meaning—the meaning it had for the artist—most historians today hold that a work has several meanings: the meaning it had for the artist, the meaning(s) it had for its first audience, the meaning(s) it had for later audiences, and the meaning(s) it has for us today. Michelangelo's *David* (page 28), for instance, in sixteenth-century Florence seems to have "meant" freedom from tyranny—the Florentines twice drove out the Medici and established republics—but most of today's viewers, unaware of the history of Florence, do not find this meaning in it.

Similarly, a portrait by Sir Joshua Reynolds (1723–92) meant one thing to the sitter's descendants when they viewed it in their ancestral country house, and it means something else to us when we view it in a museum. Picasso offers a relevant comment about changes in meaning:

> A picture is not thought out and settled beforehand. While it is being done it changes as one's thoughts change. And when it is finished, it still goes on changing, according to the state of mind of whoever is looking at it. A picture lives a life like a living creature, undergoing the changes imposed on us by our life from day to day. This is natural enough, as the picture lives only through the man who is looking at it.
>
> —Conversation with Christian Zervos, 1935, reprinted in
> *Picasso on Art,* ed. Dore Ashton (1972), 8

Although viewers usually agree in identifying the subject matter of a work of art (the martyrdom of St. Catherine, a portrait of Napoleon, a bowl of apples), disputes about subject matter are not unknown. Earlier in the chapter, for example, we saw that one of Rembrandt's paintings can be identified as *Self-Portrait with Saskia* or—a very different subject—as *The Prodigal Son;* similarly, later in Chapter 4 we will see that Rembrandt's painting of a man holding a knife has been variously identified as *The Butcher, The Assassin,* and *St, Bartholomew.* Of course an interpretation usually goes further than identifying the subject. We have already seen that Kenneth Clark interprets the picture with Saskia not only as an illustration of the story of the Prodigal Son but also as Rembrandt's expression of an insight into his relationship with his wife. Similarly, an interpretation may begin by saying that Millet's picture shows poor women gleaning, and it may go on to argue that it shows (or asserts, or represents, or expresses) some sort of theme, such as the dignity of labor, or the oppression of the worker, or the bounty of nature, or whatever.

Who Creates "Meaning"—Artist or Viewer?

Artists themselves sometimes offer interpretations of their works. For instance, writing of his *Night Café* (1888, Yale University Art Gallery), van Gogh said in a letter (8 September 1888):

> The room is blood-red and dark yellow with a green table in the middle; there are four lemon-yellow lamps with a glow of orange and green. Everywhere there is a clash and contrast of the most disparate reds and greens . . . in the empty, dreary room. . . . I have tried to express the idea that the café is a place where one can ruin oneself, run mad, or commit a crime. So I have tried to express, as it were, the powers of darkness. . . .

Many viewers find this comment helpful, but what do we make of his comment (in a letter to Gauguin, in October 1888) that the picture of his bedroom expresses an "absolute restfulness" and (in a letter to his brother, Theo, in October 1888) that the "color is to . . . be suggestive here of *rest* or of sleep in general"? Probably most viewers find the heightened perspective and the bright red coverlet on the bed disturbing rather than restful. (In the letter to Gauguin, van Gogh himself speaks of the coverlet as blood-red.) Does the artist's **intention** limit the meaning of a work? (Earlier in this chapter we touched on intention: If Zuni creators of war god figures did not intend them to be works of art, can we properly consider these creations to be works of art?) Surely one can argue that the creators of artworks may not always be consciously aware of what they are including in the works. And in stating their views they may even be consciously deceptive. Roy Lichtenstein told an interviewer, "I wouldn't believe anything I tell you."

Some modern critical theory holds that to accept the artist's statement about what he or she intended is to give the artist's intention an undeserved status. In this view, a work is created not by an isolated genius—the isolated genius is said to be a romantic invention—but by the political, economic, social, and religious ideas of a society that uses the author or artist as a conduit. Most obviously, artists such as Rembrandt, Rubens, and Titian (and in our own time, Andy Warhol, who presided over a site of production called The Factory) worked with circles of assistants and apprentices, and provided objects that responded to the demands of the market. The idea that the creator of the work cannot comment definitively on it is especially associated with Roland Barthes (1915–80), author of "The Death of the Author," in *Image-Music-Text* (1977), and with Michel Foucault (1926–84), author of "What Is an Author?" in *The Foucault Reader* (1984). For instance, in "What Is an Author?" Foucault assumes that the concept of the author (we can say the artist) is a repressive invention designed to impede the free circulation of ideas. In Foucault's view, the work does not belong to the alleged maker, who, to repeat, is a conduit for the ideas of the period. Rather, the work belongs—or ought to belong—to the *perceivers*, who of course interpret it variously, according to their historical, social, and psychological state. Works of art, James Elkins argues in *Critical Inquiry* 22 (1996), page 591, have nothing to say except what we say to them. This view, called **reception theory,** holds that art is not a body of works but is, rather, an activity of perceivers making sense of images. A work does not have meaning "in itself"; it can only mean something to someone in a context.

✍ A RULE FOR WRITERS:

Because most artists have not told us of their intentions, and because even those artists (or patrons or agents) who have stated their intentions may not be fully reliable sources, and because we inevitably see things from our own points of view, *think twice before you attribute intention to the artist* in statements such as "The designers of the stained glass windows at Chartres were trying to show us . . . ," or "Mary Cassatt in this print is aiming for . . . ," or "In his most recent photographs Hiroshi Sugimoto seeks to convey. . . ."

If one agrees that the beholders make or create the meaning, one can easily dismiss the statements that artists make about the meaning of their work. For example, although Georgia O'Keeffe on several occasions insisted that her paintings of calla lilies and of cannas were not symbolic of human sexual organs, we can (some theorists hold) ignore her comments. If *we* see O'Keeffe's lilies with their prominent stamens as phallic, and her cannas as vulval, that *is* their meaning—for us. One curator forcefully expresses this side of a highly debatable idea (although he is speaking of pictures, presumably he would extend his remarks to other kinds of artworks):

> Pictures do not have meanings. They are given meanings, by people. Different people give them different meanings at different times. One cannot merely examine a painting and try to deduce its "meaning," for such a meaning does not exist. Meaning can only ever be the outcome of a particular set of historical circumstances, and since these circumstances change, a painting's meaning cannot remain a fixed constant.
> —Paul Taylor, "Looking and Overlooking," *Art History* 15 (1992):108

Much can be said on behalf of this idea—and much can be said (and in later pages *will* be said) against it. On its behalf one can say, first, that we can never reconstruct the artist's intentions and sensations. Van Gogh's *Sunflowers*, or his portrait of his physician, can never mean for us what they meant for van Gogh. Second, the boundaries of the artwork, it is said, are not finite. The work is not simply something "out there," made up of its own internal relationships, independent of a context (*decontextualized* is the term now used). Rather, the artwork is something whose internal relationships are supplemented by what is outside of it—in the case of van Gogh, by a context consisting of the artist's personal responses to flowers and to people, and by his responses to other pictures of flowers and people, and by our responses to all sorts of related paintings, and (to give still another example) by our understanding of van Gogh's place in the history of art. Because we now know something of his life and something of the posthumous history of his paintings, we cannot experience his work in the same way or ways—in the same context—that its original audience experienced it.

When We Look, Do We See a Masterpiece—or Ourselves?

Writing an essay of any kind ought not to be an activity that you doggedly engage in to please an instructor; rather, it ought to be a stimulating, if taxing, activity that educates you and your reader. The job is twofold—seeing and saying—because these two activities are inseparable. If you don't see clearly, you won't say anything interesting and convincing. In any case, if you don't write clearly, your reader won't see what you have seen, and perhaps you haven't seen it clearly either. What you say, in short, is a device for helping the reader *and yourself* to see clearly.

But what do we see? It is now widely acknowledged that when we look, we are not looking objectively, looking with what has been called an **innocent eye.** That is, we are not like the child who, uncorrupted by the ways of fawning courtiers, accurately saw that the emperor was wearing no clothes. Inevitably, we see from a particular point of view (even if we are not aware of it)—for instance, the view of an aging middle-class white male, or of a second generation Chinese-American, or of a young Chicana feminist in the early years of the twenty-first century. Our interpretations of experience certainly feel like our own, but, far from being objective, they are (it is widely believed) largely conditioned by who we are—and who we are depends partly on the cultures that have shaped us.

Most people would probably agree with the philosopher Nelson Goodman, who in *The Languages of Art* (1968) says that what the eye sees "is regulated by need and prejudice. It selects, rejects, organizes, discriminates, associates, classifies, analyzes, constructs. It does not so much mirror as take and make" (pages 7–8). Some recent critics, influenced by Claude Lévi-Strauss's *The Savage Mind* (1966), have pushed this idea even further: Perceiving and interpreting are, they say, a form of **bricolage** (from the French *bricole,* meaning "trifle")—a form of spontaneously creating something new by assembling bits and pieces of whatever happens to be at hand or, in this case, whatever happens to be in the viewer's mind.

Mark Tansey, *The Innocent Eye Test,* 1981. Oil on canvas, 78″ × 120″. Tansey's cow—an "innocent eye"—is looking at Tansey's version of a life-size painting, Paulus Potter's *The Young Bull* (1647), in Mauritshuis, The Hague. To the right is one of Monet's paintings of grainstacks, another motif the cow is expected to be interested in. Scientists—including the one at the left equipped with a mop—will record the cow's response. (The Metropolitan Museum of Art, promised gift of Charles Cowles, in honor of William S. Lieberman, 1988; courtesy of Curt Marcus Gallery, New York)

These ideas have engendered distrust of the traditional concepts of *meaning, genius,* and *masterpiece.* The arguments, offered by scholars who belong to a school of thought called the *New Historicism,* run along these lines: Works of art are not the embodiments of profound meanings set forth by individual geniuses; rather, works of art are the embodiments of the ideology of the age that produced them. To talk of genius is to fetishize the individual. Works of art, in this view, are created not so much by exceptional individuals as by the "social energies" of a period, which somehow find a conduit in a particular artist. The old idea of a masterpiece—a work demonstrating a rare degree of skill, embodying a profound meaning, and exerting a universal appeal—thus is called into question. Theorists of the New Historicism argue that to believe in masterpieces is to believe, wrongly, that a work of art embodies an individual artist's fixed, transcendent achievement, whereas in fact (they argue) the work originally embodied the politics of the artist's age and it is now interpreted by the politics of the viewer's age. According to this way of thinking, the **canon**—the body of artworks that supposedly have stood the test of time because of their inherent quality—is not in fact a body of work of inherently superior value but is largely a construction made for political reasons by a self-serving elite. Thus, eighteenth-century landscapes of country estates with ploughed fields or with grazing cattle, it is argued, are in effect propaganda on behalf of landowners, intended to suggest that the landowners are benevolent stewards of their property. Or to take an even more obvious case, we can think about a body of work that until recently was regularly excluded from the canon: The work of women artists has been scandalously neglected because (again, in the view of some writers) patriarchal values have determined the canon. (For additional remarks about analyses that see art as material that does "cultural work," see the discussion of cultural materialism on pages 153–54.)

The idea of a universal appeal thus is said to be a myth created by a coterie (chiefly dead white males) that has succeeded in imposing its tastes and values on the rest of the world. According to these historians, the claim that, say, ancient Greece produced masterpieces of universal appeal, with the implication that all people *should* feel uplifted or enlightened or moved by these works of genius, is the propaganda of European colonialism. In this view, individualism—the idea underlying the cult of genius—is merely another bourgeois value.

But the matter need not be put so bluntly, so crudely. We can hardly doubt that our perceptions are influenced by who we are, but we need not therefore speak dismissively of our perceptions or of the objects in front of us. True, talk about "universal appeal" is a bit highfalutin, but some works of art have so deeply interested so many people over so many years that we should hesitate before we dismiss these objects as nothing but the expression of the values of a particular class.

The Relevance of Context: The Effect of the Museum

We can look at ancient Greek sculptures or at Olmec sculptures in a museum—or at pictures of them in a book—but we cannot experience them as the Greeks or the Olmecs did, in their social and religious contexts. (Indeed, most of the Greek sculptures that we see today are missing limbs or heads and have lost their original color, so we aren't really looking at what the Greeks looked at.) Historians may think that they can recreate the context—the requirements of the patrons, the studio conditions of the sculptors, the religious beliefs of the viewers, and the churches or temples in which the objects were situated—but inevitably the historians (or, rather, all of us) unwittingly project current attitudes into a constructed past. Norman Bryson and Mieke Bal, in "Semiotics and Art History," *Art Bulletin* 73 (1991), put it this way: "What we take to be positive knowledge is the product of interpretive choices. The art historian is always present in the construct she or he produces" (page 175). We cannot even become mid-twentieth-century Americans contemplating American paintings whose meaning in part was in their apparently revolutionary departure from European work. (Even those of the original viewers who are still living now see the works somewhat differently from the way they saw them in the 1950s.)

Meaning, the argument goes, is indeterminate. Further, one can add that when a museum decontextualizes the work, or deprives it of its original context—for instance, by presenting on a white wall an African mask that once was worn by a dancer in an open place, or by presenting in a vitrine with pinpoint lighting a Japanese tea bowl that had once passed from hand to hand in a humble tea house—the museum thereby invites the perceivers to project their own conceptions onto the work. A well-intended liberal effort to present Chicano art in an art museum met with opposition from the radical left, which said that the proposed exhibition was an attempt to depoliticize the works and to appropriate them into bourgeois culture. In other words, it was argued that by framing (so to speak) the works in a museum rather than in their storefront context, the works were drained of their political significance and were turned into art—mere aesthetic objects in a museum. The frame (the context) is not neutral; rather, it becomes part of what it frames.

Much of what has been said about "white box" museum displays, with their implication that museums are repositories of timeless values that transcend cultural boundaries, also can be said about the illustrations of art objects in books. Here works of art are presented (at least for the most part) in an aesthetic context, rather than in a social context of, say, economic and political forces. Indeed, we have already seen that some objects—Zuni war god figures—are sometimes taken out of their cultural context and then are presented (by a sort of benevolent colonialism, it is said) as possessing a new value: artistic merit. Some critics argue that to take a non-Western object out of its cultural context and to regard it as an independent work of art by discussing it in aesthetic terms is itself a Eurocentric (Western) colonial assault on the other culture, a denial of that culture's unique identity.

Conversely, it has been objected, when a book or a museum takes a single art object and surrounds it with abundant information about the cultural context, it demeans the object, reducing it to a mere cultural artifact—something lacking inherent value, something interesting only as part of a culture that is "the Other," remote and ultimately unknowable. Fifty years ago it was common for art historians to call attention to the aesthetic properties within a work, and for anthropologists to try to tell us "the meaning" of a work; today it is common for art historians to borrow ideas from a new breed of anthropologists, who tell us that we can never grasp the meaning of an object from another culture, and that we can understand only what it means in *our* culture. That is, we study it to learn what economic forces caused us to wrest the work from its place of origin, and what psychological forces cause us to display it on our walls. The battle between, on the one hand, providing a detailed context (and thus perhaps suggesting that the work is alien, "Other") and, on the other hand, decontextualizing (and thus slicing away meanings that the work possessed in its own culture) is still going on.°

Arguing an Interpretation (Supporting a Thesis)

Against the idea that works of art have no inherent core of meaning, and that what viewers see depends on their class or gender or whatever, one can argue that competent artists shape their work so that their intentions or meanings are evident to competent viewers (perhaps after some historical research). Most people who write about art make this assumption, and indeed such a position strikes most people as being supported by common sense.

It should be mentioned, too, that even the most vigorous advocates of the idea that meaning is indeterminate do not believe that all discussions of art are equally significant. Rather, they usually agree that a discussion is offered against a background of ideas—shared by writer and reader—as to what constitutes an effective argument, an effective presentation of a thesis. (As we saw on page 9, Kenneth Clark's thesis—or, because his thesis is tentative, we can call it a hypothesis—is that Rembrandt's *Self-Portrait with Saskia* "may express some psychological need in Rembrandt to reveal his discovery that he and his wife were two very

---

°For online reviews of exhibitions, see
CAA.reviews, ⟨*www.caareviews.org*⟩.

✍ A RULE FOR WRITERS:

Support your thesis—your point—with evidence.

different characters." Similarly, as we noted on page 13, Robert Herbert's thesis is that Millet's *The Gleaners* celebrates the heroic nature of the peasants.) When good writers offer a thesis, they do so in an essay that is

- **plausible** (reasonable because the thesis is supported with evidence)
- **coherent** (because it is clearly and reasonably organized)
- **rhetorically effective** (for instance, the language is appropriate to the reader; technical terms are defined if the imagined audience does not consist of specialists)

This means that the writer cannot merely set down random expressions of feeling or even of unsupported opinions. To the contrary, the writer tries to persuade us by *arguing* a case—by pointing to evidence that causes us to say, in effect, "Yes, I see just what you mean, and what you say makes sense."

As readers, when do we say to ourselves, "Yes, this makes sense"? And what makes us believe that one interpretation is better than another? Probably the interpretations that make sense and that strike us as better than other interpretations are the ones that are more inclusive; they are more convincing because they account for more details of the work. The less sensible, less satisfactory, less persuasive interpretations of the supposed meaning(s) are less inclusive; they leave a reader pointing to some aspects of the work—to some parts of the whole—and saying, "Yes, but this explanation doesn't take account of . . ." or "This explanation is in part contradicted by . . . ."

We'll return to the problem of interpreting meaning when we consider the distinction between subject matter and content in Chapter 2 (pages 31–32). For now, we should keep in mind two things. The first is E. H. Gombrich's comment that to a person who has been waiting for a bus, every distant object looks like a bus. The second is the implication within this comment: After the initial mistake, the person recognizes the object for what it really is.

EXPRESSING OPINIONS: THE WRITER'S "I"

The study of art is not a science, but neither is it the expression of random feelings loosely attached to works of art. You can—and must—come up with statements that seem true to the work itself, statements that almost seem self-evident (like Clark's words about Rembrandt) when the reader of the essay turns to look again at the object.

Of course works of art evoke emotions—not only nudes, but also, for example, the sprawled corpse of a rabbit in a still life by Chardin, or even the jagged edges or curved lines in a nonobjective painting. It is usually advisable, however, to reveal your feelings not by continually saying "I feel" and "this moves me," but by pointing to evidence, by calling attention to qualities in the object that shape your feelings. Thus, if you are writing about Picasso's *Les Demoiselles d'Avignon* (see page 26), instead of saying, "My first feeling is one of violence and unrest," it is better to call attention (as John Golding does, in *Cubism*) to "the savagery of the two figures at the right-hand side of the painting, which is accentuated by the lack of expression in the faces of the other figures." Golding cites this evidence in order to support his assertion that "the first impression made by the *Demoiselles* . . . is one of violence and unrest." The point, then, is not to repress or to disguise one's personal response but to account for it and to suggest that it is not eccentric and private. Golding can safely assume that his response is tied to the object and that we share his initial response because he cites evidence that compels us to feel as he does— or at least evidence that explains why we feel this way. Here, as in most effective criticism, we get what has been called "persuasive description." It is persuasive largely because it points to evidence, but also because most of us have been taught—rightly or wrongly—to respect the authority of an apparently detached point of view.

Most instructors probably would rather be alerted to the evidence in the work of art than be informed about the writer's feelings, but to say that a writer should not keep repeating "I feel" is not to say that "I" cannot be used. Nothing is wrong with occasionally using "I," and noticeable avoidances of it—"it is seen that," "this writer," "the author," "we," and the like—suggest an offensive sham modesty. Further, in accordance with the current view that we *cannot* be objective, some writers think it is only fair to inform the reader of their biases, and of how they came to hold their present views. But does highly confessional writing work, or does it turn the reader off? Most of us who read about art want to learn about art, not about the writer's life, and we may even feel that abundant

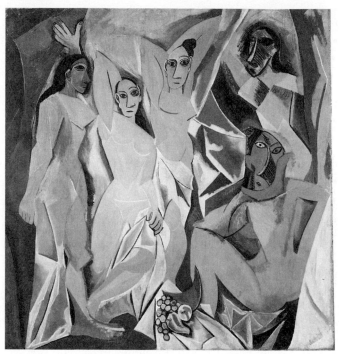

Pablo Picasso, *Les Demoiselles d'Avignon.* Paris (June-July 1907). Oil on canvas, 8' × 7'8"
(243.9 × 233.7 cm). The Museum of Modern Art, New York. Acquired through the Lillie
P. Bliss Bequest. Photograph © 1999 The Museum of Modern Art, New York. © 1999
Estate of Pablo Picasso/Artists Rights Society/ARS, New York.

autobiographical passages are excrescences. Like someone else's pen-
cilled notes in library books, abundant personal annotations are likely to
strike a reader as arrogant and unintentionally trivial.

Finally, it must be admitted that the preceding paragraphs make it
sound as if writing about art is a decorous business. In fact, it is often a
loud, contentious business, filled with strong statements about the de-
cline of culture, revolution, pornography (or a liberating sexuality), the
destruction of the skyline, fraud, new ways of seeing, and so forth. Examining the conflicting critical assumptions and methodologies will be part
of your education, and if you find yourself puzzled you will also find yourself stimulated. An energetic conversation about art has been going on
for a long time, and it is now your turn to say something.

2

Analysis

ANALYTIC THINKING: SEEING AND SAYING

An **analysis** is, literally, a separating into parts in order to understand the whole. When you analyze, you are seeking to account for your experience of the work. (Analysis thus includes **synthesis,** the combination of the parts into the whole.) You might, for example, analyze Michelangelo's marble statue *David* (see page 28) by considering:

- Its sources (in the Bible, in Hellenistic sculpture, in Donatello's bronze *David,* and in the political and social ideas of the age— e.g., David as a civic hero, the enemy of tyranny, and David as the embodiment of Fortitude)
- Its material and the limitations of that material (marble lends itself to certain postures but not to others, and marble has an effect—in texture and color—that granite or bronze or wood does not have)
- Its pose (which gives it its outline, its masses, and its enclosed spaces or lack of them)
- Its facial expression
- Its nudity (a nude Adam is easily understandable, but why a nude David?)
- Its size (here, in this over-life-size figure, man as hero)
- Its context, especially its site in the sixteenth century (today it stands in the rotunda of the Academy of Fine Arts, but in 1504 it stood at the entrance to the Palazzo Vecchio—the town hall— where it embodied the principle of the citizen-warrior and signified the victory of republicanism over tyranny)

and anything else you think the sculpture consists of—or does not consist of, for Michelangelo, unlike his predecessor Donatello, does not include the head of the slain Goliath, and thus Michelangelo's image is not explicitly that of a conquering hero. Or you might confine your attention to any one of these elements.

27

Michelangelo, *David*, 1501–1504. Marble, 13'5". Accademia, Florence. (Alinari/Art Resource, NY)

Analysis is not a process used only in talking about art. It is commonly applied in thinking about almost any complex matter. Martina Hingis plays a deadly game of tennis. What makes it so good? How does her backhand contribute? What does her serve do to her opponents? The relevance of such questions is clear. Similarly, it makes sense, when you are writing about art, to try to see the components of the work.

Here is a very short analysis of one aspect of Michelangelo's painting *The Creation of Adam* (1508–12) on the ceiling of the Sistine Chapel (see page 29). The writer's *thesis,* or the point that underlies his analysis, is, first, that the lines of a pattern say something, communicate something to the viewer, and, second, that the viewer does not merely *see* the pattern but also experiences it, participates in it.

Michelangelo, *The Creation of Adam*, 1508–12. Fresco, 9′2″ × 18′8″. Sistine Chapel, Vatican City. (Alinari; The Vatican Collection, Rome/Art Resource, NY)

The "story" of Michelangelo's *Creation of Adam,* on the ceiling of the Sistine Chapel in Rome, is understood by every reader of the book of Genesis. But even the story is modified in a way that makes it more comprehensible and impressive to the eye. The Creator, instead of breathing a living soul into the body of clay—a motif not easily translatable into an expressive pattern—reaches out toward the arm of Adam as though an animating spark, leaping from fingertip to fingertip, were transmitted from the maker to the creature. The bridge of the arm visually connects two separate worlds: the self-contained compactness of the mantle that encloses God and is given forward motion by the diagonal of his body; and the incomplete, flat slice of the earth, whose passivity is expressed in the backward slant of its contour. There is passivity also in the concave curve over which the body of Adam is molded. It is lying on the ground and enabled partly to rise by the attractive power of the approaching creator. The desire and potential capacity to get up

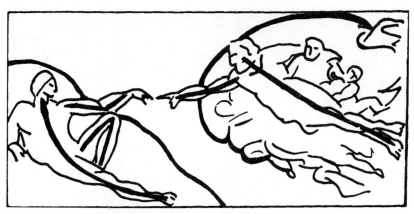

Rudolf Arnheim, diagram of Michelangelo's *Creation.* (Rudolf Arnheim, *Art and Visual Perception: A Psychology of the Creative Eye.* Copyright © 1954 The Regents of the University of California (University of California Press, 1974), pp. 458–460.)

and walk are indicated as a subordinate theme in the left leg, which also serves as a support of Adam's arm, unable to maintain itself freely like the energy-charged arm of God.

Our analysis shows that the ultimate theme of the image, the idea of creation, is conveyed by what strikes the eye first and continues to organize the composition as we examine its details. The structural skeleton reveals the dynamic theme of the story. And since the pattern of transmitted, life-giving energy is not simply recorded by the sense of vision but presumably arouses in the mind a corresponding configuration of forces, the observer's reaction is more than a mere taking cognizance of an external object. The forces that characterize the meaning of the story come alive in the observer and produce the kind of stirring participation that distinguishes artistic experience from the detached acceptance of information.

—Rudolf Arnheim, *Art and Visual Perception* (1974), 458–60

Notice that Arnheim does not discuss color, or the Renaissance background, or the place of the work in its site or in Michelangelo's development, though any or all of these are fit topics also. He has chosen to analyze the effect of only one element, but his paragraphs *are* an analysis, an attempt to record perceptions and to reflect on them.

SUBJECT MATTER AND CONTENT

Before we go on to analyze some of the ways in which art communicates, we can take a moment to distinguish between the *subject matter* of a work and the *content* or *meaning*. (Later in this chapter, on pages 85–87, we will see that the content or meaning is expressed through the *style* or *form*.)

The study of artistic images and the cultural thoughts and attitudes that they reflect is called iconology (see pages 166–71). Two pictures of the same subject matter—for instance, the Crucifixion—can express different meanings: One picture can show Christ's painful death (head drooping to one side, eyes closed, brows and mouth contorted, arms pulled into a V by the weight of the body, body twisted into an S shape); the other can show Christ's conquest of death (eyes open, face composed, arms horizontal, body relatively straight and self-possessed). The subject matter in both is the same—the Crucifixion—but the meaning or content (painful death in one picture, the conquest of death in the other) is utterly different. (The image of Christ Triumphant was common in the twelfth and early thirteenth centuries; the Suffering Christ, emphasizing the mortal aspect of Jesus, was common in the later thirteenth and the fourteenth centuries.)

To turn to another genre, if we look at some nineteenth-century landscapes we may see (aided by Barbara Novak's *Nature and Culture: American Landscape and Painting, 1825–1875*) that the *subject matter* of skies streaked with red and yellow embodies a *content* that can be described, at least roughly, as the grandeur of God. Perhaps Paul Klee was trying to turn our attention from subject matter to content when he said, "Art does not reproduce the visible; rather, it makes visible," or (in a somewhat freer translation), "Art does not reproduce what we see; rather, it makes us see."

The content, one might say, is the subject matter transformed or recreated or infused by intellect and feeling with meaning—in short, the content is a meaning made visible. This is what Henri Matisse was getting at when he said that drawing is "not an exercise of particular dexterity but above all a means of expressing intimate feelings and moods."

Even abstract and nonobjective works of art probably make visible the artist's inner experiences and thus have a subject matter that conveys a meaning. Consider Picasso's words:

> There is no abstract art. You must always start with something. Afterward you can remove all traces of reality. There is no danger then, anyway, because the idea of the object will have left an indelible mark. It is

what started the artist off, excited his ideas, and stirred up his emotions. Ideas and emotions will in the end be prisoners in his work.

—*Picasso on Art,* ed. Dore Ashton (1972), 64

This seems thoroughly acceptable. Perhaps less acceptable at first, but certainly worth pondering, is Wassily Kandinsky's remark: "The impact of an acute triangle on a sphere generates as much emotional impact as the meeting of God and Adam in Michelangelo's *Creation.*" In this exaggeration Kandinsky touches on the truth that a painting conveys more than the objects that it represents. Still, lest we go too far in searching for a content in or behind or under the subject matter, we should recall a story. In the 1920s the poet Paul Eluard was eloquently talking to Joan Miró about what Eluard took to be a solar symbol in one of Miró's paintings. After a decent interval Miró replied, "That's not a solar symbol. It is a potato."

FORM AND CONTENT

The meaning or content of a work of art is not the opposite of form. To the contrary, the *form*—including such things as the size of the work, the kinds of brush strokes in a painting, and the surface texture of a sculpture—is part of the meaning. For example, a picture with short, choppy, angular lines will "say" something different from a picture with gentle curves, even though the subject matter (let's say a woman sitting at a table) is approximately the same. When Klee spoke of "going for a walk with a line," he had in mind a line's ability (so to speak) to move quickly or slowly, assertively or tentatively. Of course many of the words we use in talking about lines—or shapes or colors—are metaphoric. If, for instance, we say that a line is "agitated" or "nervous" or "tentative" or "bold" we are not implying that the line is literally alive and endowed with feelings. We are really talking about the way in which we perceive the line, or, more precisely, we are setting forth our inference about what the artist intended or in fact produced, but such talk is entirely legitimate.

Are the lines of a drawing thick or thin, broken or unbroken? A soft pencil drawing on pale gray paper will say something different from a pen drawing made with a relatively stiff reed nib on bright white paper; at the very least, the medium and the subdued contrast of the one are quieter than those of the other. Similarly, a painting with a rough surface built up with vigorous or agitated brush strokes will not say the same thing—and will not have the same meaning—as a painting with a smooth, polished surface that gives no evidence of the brush. If nothing else, the painting that gives evidence of brush strokes announces the presence of the painter, whereas the polished surface seems to eliminate the painter from the painting.

For obvious examples, compare a work by an Action painter of the late 1940s and the 1950s such as Jackson Pollock (as you can see from the illustration on page 269, the marks on the canvas almost let us see the painter in the *act* of brushing or dribbling or spattering the pigment) with a work by a Pop artist such as Andy Warhol or Robert Indiana. Whereas Pollock executed apparently free, spontaneous, self-expressive, nonfigurative pictures, Pop artists tended to favor commonplace images (e.g., Warhol's Campbell's soup cans) and impersonal media such as the serigraph. Their works call to mind not the individual artist but anonymous commercial art and the machine, and these commercial, mechanical associations are part of the meaning of the works. Such works express what Warhol said in 1968: "The reason I'm painting this way is because I want to be a machine."

In short, to get at the content or meanings of a work we have to interpret the subject matter, the material and the form (size, shape, texture, color, and the like), the sociohistoric content, and (if known) perhaps the artist's intentions. We also have to recognize that our own sociohistoric context—including our gender, economic background, political convictions, and so forth—will to some degree determine the meanings we see in a work. Nelson Goodman, you may recall from Chapter 1 (page 19), says that because the perceiver's eye "is regulated by need and prejudice" the eye "does not so much mirror as take and make." And in our discussion (page 18) of Paul Taylor's assertion that works of art do not have meanings, we encountered an extreme version of this position, the claim that all interpretations—all discussions of content—are misinterpretations. One also hears that no standards (e.g., common sense, or the artist's intention) can guide us in evaluating different interpretations.

GETTING IDEAS: ASKING QUESTIONS TO GET ANSWERS

The painter Ad Reinhardt once said that "Looking is not as simple as it looks." Not until one has learned to look at art can one have useful ideas that one begins to set forth in writing. As Robert Frost said (with some overstatement), "All there is to writing is having ideas." What are some of the basic things to look for in trying to acquire an understanding of the languages of art—that is, in trying to understand what a work of art expresses?

Basic Questions

One can begin a discussion of the complex business of expression in the arts almost anywhere, but let's begin with some questions that can be asked of almost any work of art—whether a painting or a drawing or a sculpture or even a building.

What is my first response to the work? Amusement? Awe? Bafflement? Erotic interest? Annoyance? Shock? Boredom? Later you may modify or even reject this response, but begin by trying to study it. Jot down your responses—even your free associations. And *why* do you have this response? The act of jotting down a response, and of accounting for it analytically, may help you to deepen the response, or even to move beyond it to a different response.

When and where was the work made? By whom, and for whom? Does it reveal the qualities or values that your textbook attributes to the culture? (Don't assume that it does; works of art have a way of eluding easy generalizations.)

What does the *form* contribute? Take account of (a) *the material* (for instance, polished marble vs unpainted wood, or transparent watercolor vs opaque oil paint); (b) *the size* (a larger-than-life image will have an impact different from a miniature); (c) *the color* (realistic, or symbolic?); (d) *the composition* (balanced, or asymmetrical? highly patterned or not?).

Where would the work originally have been seen? Perhaps in a church or a palace, or a bourgeois house, or (if the work is an African mask) worn by a costumed dancer, but surely not in a textbook and not (unless it is a contemporary work) in a museum. For Picasso, "The picture-hook is the ruination of a painting. . . . As soon as [a painting] is bought and hung on a wall, it takes on quite a different significance, and the painting is done for." If the work is now part of an exhibition in a museum, how does the museum's presentation of the work affect your response?

What purpose did the work serve? To stimulate devotion? To impress the viewer with the owner's power? To enhance family pride? To teach? To delight? Does the work present a likeness, or express a feeling, or illustrate a mystery?

In what condition has the work survived? Is it exactly as it left the artist's hands, or has it been damaged, repaired, or in some way altered? What evidence of change can be seen?

What is the *title*? Does it help to illuminate the work? Sometimes it is useful to ask yourself, "What would I call the work?" Picasso called one of his early self-portraits *Yo Picasso* (i.e., "I Picasso"), rather than, say, *Portrait of the Artist,* and indeed his title goes well with the depicted self-confidence. Charles Demuth called his picture of a grain elevator in his hometown of Lancaster, Pennsylvania, *My Egypt,* a title that nicely evokes both the grandeur of the object (the silo shafts and their cap resemble an Egyptian temple) and a sense of irony (Demuth, longing to be in New York or Paris, was "in exile" in Lancaster).

Note, however, that many titles were not given to the work by the artist, and some titles are positively misleading. Rembrandt's *Night Watch* was given that name at the end of the eighteenth century, when the painting had darkened; it is really a daytime scene. And we have already noticed, on pages 8–10, that one's response to a Rembrandt painting may differ, depending on whether it is titled *Self-Portrait with Saskia* or *The Prodigal Son.*

When you ask yourself such basic questions, answers (at least tentative answers) will come to mind. In the language of today's critical theory, by means of "directed looking" you will be able to "decode" (i.e., understand) "visual statements." In short, you will have some ideas, material that you will draw on and will shape when you are called on to write. Following are additional questions to ask, first on drawing and painting, then on sculpture, architecture, photography, and video art.

Drawing and Painting

What is the **subject matter?** *Who* or *what* can we identify in the picture? What (if anything) is happening?

If the picture is a **figure painting,** what is the relation of the viewer's (and the artist's) **gaze** to the gaze of the figure(s)? After all, the viewer—the bearer of the gaze—is looking at an "Other." Does this Other return the viewer's gaze, thereby asserting his or her identity and power, or does the subject look elsewhere, unaware of the voyeur viewer-painter? It has been argued, for instance, that in his pictures of his family and friends, Degas gives his subjects a level stare, effectively placing them on the same social level as the viewer; in his pictures of working women (laundresses, dancers), he adopts a high viewpoint, literally looking down on his unaware subjects; in his pictures of prostitutes, he looks either from below or from above, gazing as a spy or voyeur might do, with unsuspecting and therefore vulnerable victims.

Concern with the "gaze," and the idea that (in art) males look actively whereas women are to-be-looked-at, was perhaps first set forth by Laura Mulvey in "Visual Pleasure and Narrative Cinema," in the journal *Screen* 16:3 (1975): 6–18, reprinted in her book *Visual and Other Pleasures* (Bloomington: University of Indiana Press, 1989). In Mary Cassatt's *Woman in Black at the Opera* (c. 1878; also called *At the Français, a Sketch*, illustrated on the cover of the present book), however, there is not so simple a dichotomy. True, the woman in the foreground is being looked at by the man in the upper left, but the woman herself is very actively looking, and she is a far more dominating figure (severe profile, dark garments, large size, angular forms) than the small and somewhat comically sprawling man who is looking at her (and in effect at us). These two figures are looking, but the person who is looking at the picture—yet another, the viewer—surely sees power as located in the woman rather than in the man.

If more than one figure is shown, what is the relation of the figures to each other?

If there is only one figure, is it related to the viewer, perhaps by the gaze or by a gesture? If the figure seems posed, do you agree with those theoreticians who say that posing is a subordination of the self to the gaze of another, and the offering of the self (perhaps provocatively or shamefully) to the viewer?

Baudelaire said that a **portrait** is "a model complicated by an artist." The old idea was that a good portrait revealed not only the face but also the inner character of the figure. The face was said to be the index of the mind; thus, for instance, an accurate portrait of King X showed his cruelty (it was written all over his face), and accurate portraits of Pope Y and of Lady Z showed, respectively, the pope's piety (or worldliness) and the lady's tenderness (or arrogance). It usually turned out, however, that the art historians who saw such traits in particular portraits already knew what traits to expect. When the portrait was of an unidentified sitter, the commentaries varied greatly.

It is now widely held that a portrait is not simply a representation of a face that reveals the inner character; a portrait is also a presentation or a construction created by the artist *and* the sitter. Sitters and artists both (so to speak) offer interpretations.

How are their interpretations conveyed? Consider such matters as these:

- How much of the figure does the artist show, and how much of the available space does the artist cause the figure to occupy? What effects are thus gained?
- What do the clothing, furnishings, accessories (swords, dogs, clocks, and so forth), background, angle of the head or posture of the head and body, and facial expression contribute to our sense of the figure's personality (intense, cool, inviting)? Is the sitter portrayed in a studio setting or in his or her own surroundings?
- Does the picture advertise the sitter's *political* importance, or does it advertise the sitter's *personal* superiority? A related way of thinking is this: Does the image present a strong sense of a social class (as is usual in portraits by Frans Hals) or a strong sense of an independent inner life (as is usual in portraits by Rembrandt)?
- If frontal, does the figure seem to face us in a godlike way, as if observing everything before it? If three-quarter, does it suggest motion, a figure engaged in the social world? If profile, is the emphasis decorative or psychological? (Generally speaking, a frontal or, especially, a three-quarter view lends itself to the rendering of a dynamic personality, perhaps even interacting in an imagined social context, whereas a profile does not—or if a profile does reveal a personality it is that of an aloof, almost unnaturally self-possessed sitter.)
- If the picture is a double portrait, does the artist reveal what it is that ties the two figures together? Do the figures look at each other? If not, what is implied by the lack of eye contact?
- Is the figure (or are the figures) allegorical (turned into representations of, say, liberty or beauty or peace or war)? Given the fact that female sitters are more often allegorized than males, do you take a given allegorical representation of a female to be an act of appropriation—a male forcing a woman into the role of "Other"?
- If the picture is a self-portrait, what image does the artist project? Van Gogh's self-portraits in which he wears a felt hat and a jacket show him as the bourgeois gentleman, whereas those in which he wears a straw hat and a peasant's blouse or smock show him as the country artist.
- It is sometimes said that every portrait is a self-portrait. (In Leonardo's formula, "the painter always paints himself." In the words of Dora Maar, Picasso's mistress in the 1930s and 1940s,

"All his portraits of me are lies. They're all Picassos. Not one is Dora Maar.") Does this portrait seem to reveal the artist in some way?

- Some extreme close-up views of faces, such as those of the contemporary photo-realist painter Chuck Close, give the viewer such an abundance of detail—hairs, pores, cracks in lips—that they might be called landscapes of faces. Do they also convey a revelation of character or of any sort of social relationship, or does this overload of detail prevent the viewer from forming an interpretation?

- Does the portrait, in fact, reveal anything at all? Looking at John Singer Sargent's portrait entitled *General Sir Ian Hamilton*, the critic Roger Fry said, "I cannot see the man for the likeness." Sargent said that he saw an animal in every sitter.

For a student's discussion of two portraits by John Singleton Copley, see page 108. For a professional art historian's discussion of Anthony Van Dyck's portrait of Charles I, see page 139. For a brief, useful survey of the topic, see Joanna Woodall's introduction to a collection of essays, *Portraiture: Facing the Subject,* edited by Joanna Woodall (Manchester: Manchester UP, 1997).

Let's now consider a **still life** (plural: *still lifes,* not *still lives*)—a depiction of inanimate objects in a restricted setting, such as a tabletop.

- What is the chief interest? Is it largely in the relationships between the shapes and the textures of the objects? Or is it in the symbolic suggestions of opulence (a Dutch seventeenth-century painting, showing a rich tablecloth on which are luxurious eating utensils and expensive foods) or, on the other hand, is the interest in humble domesticity and the benefits of moderation (a seventeenth-century Spanish painting, showing a simple wooden table on which are earthenware vessels)?

- Does it imply transience, perhaps by a burnt-out candle, or even merely by the perishable nature of the objects (food, flowers) displayed? Other common symbols of *vanitas* (Latin for "emptiness," particularly the emptiness of earthly possessions and accomplishments) are an overturned cup or bowl and a skull.

- If the picture shows a piece of bread and a glass of wine flanking a vase of flowers, can the bread and wine perhaps be eucharistic symbols, the picture as a whole representing life everlasting achieved through grace?

- Is there a contrast (and a consequent evocation of *pathos*) between the inertness and sprawl of a dead animal and its vibrant color or texture? Does the work perhaps even suggest, as some of Chardin's pictures of dead rabbits do, something close to a reminder of the crucifixion?
- Is all of this allegorizing irrelevant?

When the picture is a **landscape,** you may want to begin by asking the following questions:

- What is the relation between human beings and nature? Are the figures at ease in nature (e.g., aristocrats lounging complacently beneath the mighty oaks that symbolize their ancient power and grandeur) or are they dwarfed by it? Are they earthbound, beneath the horizon, or (because the viewpoint is low) do they stand out against the horizon and perhaps seem in touch with the heavens, or at least with open air?
- Do the natural objects in the landscape (e.g., billowy clouds, or dark clouds, or gnarled trees, or airy trees) somehow reflect the emotions of the figures?
- What does the landscape say about the society for which it was created? Even if the landscape seems realistic, it may also express political or spiritual forces. Does it, for instance, reveal an aristocrat's view of industrious, well-clad peasants toiling happily in a benevolently ordered society? Does it—literally—put the rural poor in the shade, letting the wealthy people get the light? (This view is set forth in John Barrell, *The Dark Side of the Landscape: The Rural Poor in English Painting, 1730–1840,* 1980.)

In short, a landscape painting is not just an objective presentation of earth, rocks, greenery, water, and sky. The artist presents what is now called a social construction of nature—for instance, nature as a place made hospitable by the wisdom of the landowners, or nature as an endangered part of our heritage, or nature as a world that we have lost, or nature as a place where the weary soul can find rest and nourishment. (For an analysis employing recent critical approaches, see Mark Roskill, *The Language of Landscape,* 1996.)

We have been talking about particular subjects—figure painting, still life, landscape—but other questions concern all kinds of painting and drawing. Are the **contour lines** (outlines of shapes) strong and hard, isolating each figure or object? Or are they irregular, indistinct, fusing the subjects with the surrounding space? Do the lines seem (e.g., in an Asian ink painting) calligraphic—that is, of varied thicknesses that suggest liveliness or vitality—or are the lines uniform and suggestive of painstaking care?

What does the **medium** (the substance on which the artist acted) contribute? For a drawing made with a wet medium (e.g., ink applied with a pen, or washes applied with a brush), what does the degree of absorbency of the paper contribute? Are the lines of uniform width, or do they sometimes swell and sometimes diminish, either abruptly or gradually? (Quills and steel pens are more flexible than reed pens.) For a drawing made with a dry medium (e.g., silverpoint, charcoal, chalk, or pencil), what does the smoothness or roughness of the paper contribute? (When crayon is rubbed over textured paper, bits of paper show through, suffusing the dark with light, giving vibrancy.) In any case, a drawing executed with a dry medium, such as graphite, will differ from a drawing executed with a wet medium, where the motion of the instrument must be interrupted in order to replenish the ink or paint.°

If the work is a painting, is it in **tempera** (pigment dissolved in egg, the chief medium of European painting into the late fifteenth century), which usually has a somewhat flat, dry appearance? Because the brush strokes do not fuse, tempera tends to produce forms with sharp edges—or, we might say, because it emphasizes contours it tends to produce colored drawings. Or is the painting done with **oil paint,** which (because the brush strokes fuse) is better suited than tempera to give an effect of muted light and blurred edges? Thin layers of translucent colored oil glazes can be applied so that light passing through these layers reflects from the opaque ground colors, producing a soft, radiant effect; or oil paint can be put on heavily (*impasto*), giving a rich, juicy appearance. Impasto can be applied so thickly that it stands out from the surface and catches the light. Oil paint, which lends itself to uneven, gestural, bravura

---

°For a well-illustrated, readable introduction to the physical properties of drawings, see Susan Lambert, *Reading Drawings* (New York: Pantheon, 1984). For a more detailed but somewhat drier account, see James Watrous, *The Craft of Old-Master Drawings* (Madison: University of Wisconsin Press, 1957).

handling, is thus sometimes considered more painterly than tempera, or, to reverse the matter, tempera is sometimes considered to lend itself to a more linear treatment.

Chinese, Korean, and Japanese ink painting, too, illustrates the contribution of the media. A painting on silk is usually very different from a painting on paper. Because raw silk absorbs ink and pigments, thereby diluting the strength of the line and the color, silk is usually sized (covered with a glaze or filler) to make it less absorbent, indeed, slick. If the brush moves rapidly on the sized surface, it may leave a broken line, so painters working on silk usually proceed slowly, meticulously creating the image. Painters who want spontaneous, dynamic, or blurred brushwork usually paint not on silk but on paper.

Caution: Reproductions in books usually fail to convey the texture of brush strokes. If you must work from reproductions, try to find a book that includes details (small parts of the picture), preferably enlarged.

Is the **color** (if any) imitative of appearances, or expressive, or both? (Why is the flesh of the Buddha gold? Why did Picasso use white, grays, and blacks for *Guernica,* when in fact the Spaniards bombarded the Basque town on a sunny day?) How are the colors related—for example, by bold contrasts or by gradual transitions?

The material value of a pigment—that is to say, its cost—may itself be expressive. For instance, Velázquez's lavish use of expensive ultramarine blue in his *Coronation of the Virgin* in itself signifies the importance of the subject. Ultramarine—"beyond the sea"—made of imported ground lapis lazuli, was more expensive than gold; its costliness is one reason why, like gold, it was used for some holy figures in medieval religious paintings, whereas common earth pigments were used for nondivine figures.

Vincent van Gogh, speaking of his own work, said he sought "to express the feelings of two lovers by a marriage of two complementary colors, their mixture and their oppositions, the mysterious vibrations of tones in each other's proximity . . . to express the thought behind a brow by the radiance of a bright tone against a dark ground." As this quotation may indicate, comments on the expressive value of color often seem highly subjective and perhaps unconvincing. One scholar, commenting on the yellowish green liquid in a bulbous bottle at the right of Manet's *Bar aux Folies-Bergère,* suggests that the color of the drink—probably absinthe—is oppressive. A later scholar points out that the distinctive shape of the bottle indicates that the drink is crème de menthe, not absinthe, and therefore he finds the color not at all disturbing.

Caution: It is often said that *warm colors* (red, yellow, orange) come forward and produce a sense of excitement, whereas *cool colors* (blue, green) recede and have a calming effect, but experiments have proved inconclusive; the response to color—despite clichés about seeing red or feeling blue—is highly personal, highly cultural, highly varied. Still, a few things can be said, or at least a few terms can be defined. *Hue* gives the color its name—red, orange, yellow, green, blue, violet. *Value* (also called *lightness* or *darkness, brightness*) refers to relative lightness or darkness of a hue. When white is added, the value becomes "higher"; when black is added, the value becomes "lower." The highest value is white; the lowest is black. Light gray has a higher value than dark gray. *Saturation* (also called *hue intensity*) is the strength or brightness of a hue—one red is redder than another; one yellow is paler than another. A vivid hue is of high saturation; a pale hue is of low saturation. But note that much in a color's appearance depends on context. Juxtaposed against green, red will appear redder than if juxtaposed against orange. A gray patch surrounded by white seems darker than the same shade of gray surrounded by black.

When we are armed with these terms, we can say, for example, that in his South Seas paintings Paul Gauguin used *complementary colors* (orange and blue, yellow and violet, red and green, i.e., hues that when mixed absorb almost all white light, producing a blackish hue) at their highest values, but it is harder to say what this adds up to. (Gauguin himself said that his use of complementary colors was "analogous to Oriental chants sung in a shrill voice," but one may question whether the analogy is helpful.)

For several reasons our nerve may fail when we try to talk about the effect of color. For example:

- Light and moisture cause some pigments to change over the years, and the varnish customarily applied to Old Master paintings inevitably yellows with age, altering the appearance of the original.
- The colors of a medieval altarpiece illuminated by flickering candlelight or by light entering from the yellowish translucent (not transparent) glass or colored glass of a church cannot have been perceived as the colors that we perceive in a museum, and, similarly, a painting by van Gogh done in bright daylight cannot have looked to van Gogh as it looks to us on a museum wall.

The moral? Be cautious in talking about the effect of color. Keep in mind the remark of the contemporary painter Frank Stella: "Structural analysis is a matter of describing the way the picture is organized. Color analysis would seem to be saying what you think the color does. And it seems to me that you are more likely to get an area of common agreement in the former."

What is the effect of **light** in the picture? Does it produce sharp contrasts, brightly illuminating some parts and throwing others into darkness, or does it, by means of gentle gradations, unify most or all of the parts? Does the light seem theatrical or natural, disturbing or comforting? Is light used to create symbolic highlights?

Do the objects or figures share the **space** evenly, or does one overpower another, taking most of the space or the light? What is the focus of the composition? The **composition**—the ordering of the parts into a whole by line, color, and shape—is sometimes grasped at an initial glance and at other times only after close study. Is the composition symmetrical (and perhaps therefore monumental, or quiet, or rigid and oppressive)? Is it diagonally recessive (and perhaps therefore dramatic or even melodramatic)?

Are figures harmoniously related, perhaps by a similar stance or shared action, or are they opposed, perhaps by diagonals thrusting at each other? Speaking generally—very generally—**diagonals** may suggest motion or animation or instability, except when they form a triangle resting on its base, which is a highly stable form. **Horizontal lines** suggest tranquility or stability—think of plains, or of reclining figures. **Vertical lines**—tree trunks thrusting straight up, or people standing, or upright lances as in Velázquez's *Surrender of Breda*—may suggest a more vigorous stability. **Circular lines** are often associated with motion and sometimes—perhaps especially by men—with the female body and with fertility. It is even likely that Picasso's *Still-Life on a Pedestal Table,* with its rounded forms, is, as he is reported to have called it, a "clandestine" portrait of one of his mistresses. These simple formulas, however, must be applied cautiously, for they are not always appropriate. Probably it is fair to say, nevertheless, that when a *context* is established—for instance, by means of the title of a picture—these lines may be perceived to bear these suggestions if the suggestions are appropriate.

Caution: The sequence of eye movements with which we look at a picture has little to do with the compositional pattern. That is, the eye does not move in a circle when it perceives a circular pattern. The mind, not the eye, makes the relationships. It is therefore inadvisable to say things like "The eye follows the arrow and arrives finally at the target."

Does the picture convey **depth,** that is, **recession in space?** If so, how? If not, why not? (Sometimes space is flattened—e.g., to convey a sense of otherworldliness or eternity.) Among the chief ways of indicating depth are the following:

- *Overlapping* (the nearer object overlaps the farther object)
- *Foreshortening* (as in the recruiting poster *I Want You,* where Uncle Sam's index finger, pointing at the viewer, is represented chiefly by its tip, and, indeed, the forearm is represented chiefly by a cuff and an elbow)
- *Contour hatching* (lines or brush strokes that follow the shape of the object depicted, as though a net were placed tightly over the object)
- *Shading* or *modeling* (representation of shadows on the body)
- Representation of *cast shadows*
- *Relative position from the ground line* (objects higher in the picture are conceived of as further away than those lower)
- *Perspective* (parallel lines seem to converge in the distance, and a distant object will appear smaller than a near object of the same size.° Some cultures, however, use a principle of *hierarchic scale.* In such a system a king, for instance, is depicted as bigger than a slave not because he is nearer but because he is more important; similarly, the Virgin in a nativity scene may be larger than the shepherds even though she is behind them. For an example of hierarchic scale, see the sculpture by Olwe of Ise, on page 170, where the senior queen is the largest figure, the king the second largest, and the two attendants, at the king's feet, are the smallest because they are the least important.)
- *Aerial or atmospheric perspective* (remote objects may seem— depending on the atmospheric conditions—slightly more bluish than similar near objects, and they may appear less intense in color and less sharply defined than nearer objects. In Leonardo's

---

°In the Renaissance, perspective was used chiefly to create a coherent space and to locate objects within that space, but later artists have sometimes made perspective expressive. Giorgio de Chirico, for example, often gives a distorted perspective that unnerves the viewer. Or consider van Gogh's *Bedroom at Arles.* Although van Gogh said that the picture conveyed "rest," viewers find the swift recession disturbing. Indeed, the perspective in this picture is impossible: If one continues the diagonal of the right-hand wall by extending the dark line at the base, one sees that the bed's rear right foot would be jammed into the wall.

Mona Lisa, for instance, the edges of the distant mountains are blurred. *Caution:* Aerial perspective does *not* have anything to do with a bird's-eye view.)

Does the picture present a series of planes, each parallel to the picture surface (foreground, middle ground, background), or does it, through some of the means just enumerated, present an uninterrupted extension of one plane into depth?

What is the effect of the **shape** and **size** of the work? Because, for example, most still lifes use a horizontal format, perhaps thereby suggesting restfulness, a vertical still life may seem relatively towering and monumental. Note too that a larger-than-life portrait—Chuck Close's portraits are eight or nine feet high—will produce an effect different from one eight or nine inches high. If you are working from a reproduction be sure, therefore, to ascertain the size of the original.

What is the **scale,** that is, the relative size? A face that fills a canvas will produce a different effect from a face of the same size that is drawn on a much larger canvas; probably the former will seem more expansive or more energetic, even more aggressive.

A Note on Nonobjective Painting. We have already noticed (page 32) Wassily Kandinsky's comment that "The impact of an acute triangle on a sphere generates as much emotional impact as the meeting of God and Adam in Michelangelo's *Creation.*" Kandinsky (1866–1944), particularly in his paintings and writings of 1910–14, has at least as good a title as anyone else to being called the founder of twentieth-century **nonobjective art.** Nonobjective art, unlike figurative art, depends entirely on the emotional significance of color, form, texture, size, and spatial relationships, rather than on representational forms.

The term *nonobjective art* includes **abstract expressionism**—a term especially associated with the work of New York painters in the 1950s and 1960s, such as Jackson Pollock (1912–56) and Mark Rothko (1903–70), who, deeply influenced by Kandinsky, sought to allow the unconscious to express itself. Nonobjective art is considered synonymous with **pure abstract art,** but it is *not* synonymous with "abstract art," since in most of what is generally called abstract art, forms are recognizable though simplified.

In several rather mystical writings, but especially in *Concerning the Spiritual in Art* (1910), Kandinsky advanced theories that exerted a great influence on American art after World War II. For Kandinsky, colors were something to be felt and heard. When he set out to paint, he wrote, he "let himself go. . . . Not worrying about houses or trees, I spread strips and dots of paint on the canvas with my palette knife and let them sing out as loudly as I could."

Nonobjective painting is by no means all of a piece; it includes, to consider only a few examples, not only the lyrical, highly fluid forms of Kandinsky and of Jackson Pollock but also the pronounced vertical and horizontal compositions of Piet Mondrian (1872–1944) and the bold, rough slashes of black on white of Franz Kline (1910–62), although Kline's titles sometimes invite the viewer to see the slashes as representations of the elevated railway of Kline's earlier years in New York City. Nonobjective painting is not so much a style as a philosophy of art: In their works, and in their writings and their comments, many nonobjective painters emphasized the importance of the unconscious and of chance. Their aim in general was to convey feelings with little or no representation of external forms; the work on the canvas conveyed not images of things visible in the world, but intuitions of spiritual realities. Notice that this is *not* to say that the paintings are "pure form" or that subject matter is unimportant in nonobjective art. To the contrary, the artists often insisted that their works were concerned with what really was real—the essence behind appearances—and that their works were not merely pretty decorations. Two Abstract Expressionists, Mark Rothko and Adolph Gottlieb (1903–74), emphasized this point in a letter published in the *New York Times* in 1943:

> There is no such thing as good painting about nothing. We assert that the subject is crucial and only that subject-matter is valid which is tragic and timeless. That is why we profess spiritual kinship with primitive and archaic art.
>
> —Quoted in *American Artists on Art from 1940 to 1980*,
> ed. Ellen H. Johnson (1982), 14

Similarly, Jackson Pollock, speaking in 1950 of his abstract works created in part by spattering paint and by dribbling paint from the can, insisted that the paintings were not mere displays of a novel technique and were not mere designs:

> It doesn't make much difference how the paint is put on as long as something has been said. Technique is just a means of arriving at a statement.
>
> —Quoted in *American Artists on Art from 1940 to 1980*,
> ed. Ellen H. Johnson (1982), 10

For a photograph of Pollock working with his "poured" or "drip" technique, where the lines on the canvas refer not to objects but only to the gestures that made the lines, see page 269.

The **titles** of nonobjective pictures occasionally suggest a profound content (e.g., Pollock's *Guardians of the Secret,* Rothko's *Vessels of Magic*), occasionally a more ordinary one (Pollock's *Blue Poles*), and occasionally something in between (Pollock's *Autumn Rhythm*), but one can judge a picture by its title only about as well as one can judge a book by its cover (i.e., sometimes well, sometimes not at all).

In writing about the work of nonobjective painters, you may get some help from their writings, though of course you may come to feel in some cases that the paintings do not do what the painters say they want the pictures to do. Good sources for statements by artists are *Theories of Modern Art* (1968), ed. Herschel B. Chipp; *American Artists on Art from 1940 to 1980* (1982), ed. Ellen H. Johnson; *Art in Theory: 1900–1990* (1992), ed. Charles Harrison and Paul Wood; and *Theories and Documents of Contemporary Art: A Sourcebook of Artists' Writings* (1996), ed. Kristine Stiles and Peter Selz. (In reading the comments of artists, however, it is often useful to recall Claes Oldenburg's remark that anyone who listens to an artist talk should have his eyes examined.)

Finally, here is a comment about a severely geometric nonobjective picture by Frank Stella (b. 1936) (see page 48). The picture, one of Stella's Protractor series, is 10 feet tall and 20 feet wide. Robert Rosenblum writes:

> Confronted with a characteristic example, *Tahkt-i-Sulayman I,* the eye and the mind are at first simply dumbfounded by the sheer multiplicity of springing rhythms, fluorescent Day-Glo colors, and endlessly shifting planes—all the more so, because the basic components (circles and semicircles; flat bands of unmodulated color) and the basic design (here a clear bilateral symmetry) are so lucid. But again, as always in Stella's work, the seeming economy of vocabulary is countered by the elusive complexities of the result. At first glance, the overriding pattern is of such insistent symmetrical clarity that we feel we can seize predictable principles of organization and bring to rest this visual frenzy. But Stella permits no such static resolution, for the overall symmetries of the design are contradicted by both the interlace patterns and the colors, which constantly assert their independence from any simple-minded scheme. In a surprising way, this tangle of gyrating energies, released and recaptured, provides a 1960s ruler-and-compass equivalent of the finest Pollocks, even in terms of its engulfing scale (here 20 feet wide), which imposes itself in an almost physical way upon the spectator's

Frank Stella, *Tahkt-i-Sulayman I*, 1967. Fluorescent acrylic on canvas, 10′ × 20′.
(Collection, Mr. Robert Rowan; © 1992 Frank Stella/ARS, New York)

world. In this case, the springing vaults of the arcs, some reaching as high
as four feet above one's head, turn the painting into something that verges
on the architectural, a work that might rest on the floor and be subject to
natural physical laws of load and support. Seen on this immense scale, the
thrusts and counterthrusts, the taut and perfect spanning of great spaces,
the razor-sharp interlocking of points of stress all contrive to plunge the
observer into a dizzying tour-de-force of aesthetic engineering.

—*Frank Stella* (1971), 48–49

What brief advice can be given about responding to nonobjective
painting? Perhaps only this (and here is something of a repetition of what
has already been said about representational drawings and paintings): As
you look at the work, begin with your responses to the following:

- The dynamic interplay of colors, shapes, lines, textures (of
  pigments and of the ground on which the pigments are applied)
- The size of the work (often large)
- The shape of the work (most are rectangular or square, but
  especially in the 1960s many are triangular, circular, chevron-
  shaped, diamond-shaped, and so on, with the result that, because
  they depart from the traditional shape of paintings, they seem
  almost to be objects—two-dimensional sculptures rather than
  paintings)
- The title

Later, as has been suggested, you may want to think about the picture in the context of statements made by the artist—for instance, Pollock's "My concern is with the rhythms of nature, the way the ocean moves. I work inside out, like nature." Useful sources include the four collections of comments edited by Herschel B. Chipp, Ellen H. Johnson, Charles Harrison and Paul Wood, and Kristine Stiles and Peter Selz, mentioned on page 47.

Finally, remember that making a comparison is one of the most effective ways of seeing things. How does this work differ from that work, and what is the effect of the difference?

Sculpture

For what **purpose** was this object made? To edify the faithful? To commemorate heroism? What is expressed through the representation? What, for instance, does the highly ordered, symmetrical form of *King Chefren* (also called Khafre; Egyptian, third millennium BC; see page 50) suggest about the man? What is the relationship of naturalism to idealism or abstraction? (On realism and idealism, see pages 94–100.) If the sculpture represents a deity, what ideas of divinity are expressed? If it represents a human being as a deity (e.g., Alexander the Great as Herakles, or King Chefren as the son of an Egyptian deity), how are the two qualities portrayed?

If the work is a **portrait.** some of the questions suggested earlier for painted portraits (pages 36–38) may be relevant. Consider especially whether the work presents a strong sense of an individual or, on the other hand, of a type. Paradoxically, a work may do both: Roman portraits from the first to the middle of the third century are (for the most part) highly realistic images of the faces of older men, the conservative nobility who had spent a lifetime in public office. Their grim, wrinkled faces are highly individualized, and yet these signs of age and care indicate a rather uniform type, supposedly devoted and realistic public servants who scorn the godlike posturing and feigned spontaneity of such flashy young politicians as Caesar and Pompey. That is, although the model might not in fact have been wrinkled, it apparently was a convention for a portrait bust to show signs of wear and tear, such as wrinkles, thereby indicating that the subject was a hardworking, mature leader. In other societies such signs of mortality may be removed from leaders. For instance, African portrait sculpture of leaders tends to present idealized images. Thus, in Ife bronzes from the twelfth century, rulers show a commanding stance

Egyptian, *King Chefren,* ca. 2500 BC. Diorite, 5'6". (Courtesy Hirmer Fotoarchiv, Munich/Egyptian Museum, Cairo)

and a fullness of body, whereas captives (shown in order to say something not about themselves but about their conqueror) may be represented with bulging eyes, wrinkled flesh, and bones evident beneath the skin. In keeping with the tradition of idealizing, commemorative images of elders usually show them in the prime of life.

What does the **pose** imply? Effort? Rest? Arrested motion? Authority? In the Lincoln Memorial, Lincoln sits; in the Jefferson Memorial, Jefferson stands, one foot slightly advanced. Lincoln's pose as well as his face suggest weariness, while Jefferson's pose as well as his faintly smiling face suggest confidence and action. How relevant to a given sculpture is Rodin's comment that "The body always expresses the spirit for which it is the shell"?

Are certain bodily features or forms distorted? If so, why? (In most African equestrian sculpture, the rider—usually a chief or an ancestor—dwarfs the horse, in order to indicate the rider's high status.) To what extent is the **drapery** independent of the body? Does it express or diminish the **volumes** (enclosed spaces, e.g., breasts, knees) that it covers? Does it draw attention to specific points of focus, such as the head or hands? Does it indicate bodily motion or does it provide an independent harmony? What does it contribute to whatever the work expresses? If the piece is a wall or niche sculpture, does the pattern of the drapery help to integrate the work into the façade of the architecture?

If the sculpture is a bust, what sort of **truncation** (termination of the image) has the sculptor used? Does a straight horizontal line run below the shoulders, or does the bare or draped chest end in a curve? Does the sitter's garment establish the termination? Or is the termination deliberately irregular, perhaps emphasizing the bust as a work of art rather than as a realistic reproduction of the subject?

What do the **medium** and the **techniques** by which the piece was shaped contribute?* Clay is different from stone or wood, and stone or wood can be rough or they can be polished. Would the statue of Chefren (see page 50) have the same effect if it were in clay instead of in highly polished diorite? Can one imagine Daniel Chester French's marble statue of Lincoln, in the Lincoln Memorial, done in stainless steel? What are the associations of the material? For instance, early in this century welded iron suggested heavy industry, in contrast with bronze and marble, which suggested nobility, the classical world, and great wealth.

Even more important, what is the effect of the **tactile qualities;** for example, polished wood versus terra cotta? Notice that the tactile qualities result not only from the medium but also from the **facture;** that is, the process of working on the medium with certain tools. An archaic Greek *kouros* ("youth") may have a soft, warm look not only because of the porous marble but because of traces left, even after the surface was smoothed with abrasives, of the sculptor's bronze punches and (probably) chisels.

---

*Media and techniques are lucidly discussed by Nicholas Penny in *The Materials of Sculpture* (New Haven: Yale University Press, 1994). Also of use is a brief treatment, Jane Basset and Peggy Fogelman, *Looking at European Sculpture: A Guide to Technical Terms* (Los Angeles: Getty Museum, 1997).

Consider especially the distinction between **carving,** which is subtractive, and **modeling,** which is additive; that is, the difference between cutting away, to release the figure from the stone, wood, or ivory, and, on the other hand, building up or modeling, to create the figure out of a pliable material such as lumps of clay, wax, or plaster.* Rodin's *Walking Man* (see page 143), built up by modeling clay and then cast in bronze, recalls in every square inch of the light-catching surface a sense of the energy that is expressed by the figure. Can one imagine Michelangelo's *David* (see page 28), carved in marble, with a similar surface? Even assuming that a chisel could imitate the effects of modeling, would the surface thus produced catch the light as Rodin's does? And would such a surface suit the pose and the facial expression of *David?*

Compare *King Chefren* with Giovanni da Bologna's *Mercury* (see page 53). *King Chefren* was carved; the sculptor, so to speak, cut away from the block everything that did not look like Chefren. *Mercury* was modeled—built up—in clay or wax, and then cast in bronze. The massiveness or stability of *King Chefren* partakes of the solidity of stone, whereas the elegant motion of *Mercury* suggests the pliability of clay, wax, and bronze.

What kinds of **volumes** are we looking at? Geometric (e.g., cubical, spherical) or irregular? Is the **silhouette** (outline) open or closed? In Michelangelo's *David,* David's right side is said to be closed because his arm is extended downward and inward; his left side is said to be open because the upper arm moves outward and the lower arm is elevated toward the shoulder. Still, although the form of *David* is relatively closed, the open spaces—especially the space between the legs—emphasize the potential expansion or motion of the figure. The unpierced, thoroughly closed form of *King Chefren,* in contrast to the open form of *Mercury,* implies stability and permanence.

What is the effect of **color,** either of the material or of gilding or paint? Is color used for realism or for symbolism? Why, for example, in the tomb of Urban VIII, did Gian Lorenzo Bernini use bronze for the sarcophagus (coffin), the pope, and Death, but white marble for the figures of Charity and Justice? The whiteness of classical stone sculpture is usually regarded as suggesting idealized form (though in fact the Greeks tinted

---

*"Modeling" is also used to refer to the treatment of volumes in a sculpture. Deep modeling, characterized by conspicuous projections and recesses, for instance in drapery, creates strong contrasts in highlights and shadows. On the other hand, shallow modeling creates a relatively unified surface.

Giovanni da Bologna,
Mercury, 1580.
Bronze, 69″. (Alinari;
National Museum,
Florence/Art
Resource, NY)

the stone and painted in the eyes), but what is the effect of the whiteness
of George Segal's plaster casts (see page 54) of ordinary figures in ordinary
situations, in this instance of a man sitting on a real stool and a woman
standing beneath a real fluorescent light and behind a real counter, set off
by a deep-red panel at the back wall? Blankness? Melancholy?

What was the original **location** or **site** or physical context (e.g., a
pediment, a niche, a public square)?

George Segal, *The Diner*, 1964–66. Plaster, wood, chrome, laminated plastic masonite, and fluorescent lamp, 93¾" × 144¼ × 96". Collection Walker Art Center, Minneapolis; gift of the T. B. Walker Foundation, 1966. © George Segal/Licensed by VAGA, New York, NY.

Is the **base** a part of the sculpture (e.g., rocks, or a tree trunk that helps to support the figure), and, if so, is it expressive as well as functional? George Grey Barnard's *Lincoln—the Man*, a bronze figure in a park in Cincinnati, stands not on the tall classical pedestal commonly used for public monuments but on a low boulder—a real one, not a bronze copy—emphasizing Lincoln's accessibility, his down-to-earthness. Almost at the other extreme, the flying *Mercury* (see page 53) stands tiptoe on a gust of wind, and at the very extreme, Marino Marini's *Juggler* is suspended above the base, emphasizing the subject's airy skill.

Where is the best place (or where are the best places) to stand in order to experience the work? Do you think that the sculpture is intended to be seen from multiple views, all of which are equally interesting and important? Or is the work strongly oriented toward a single viewpoint, as is the case with a sculpture set within a deep niche? If so, are frontality, rigidity, and stasis important parts of the meaning? Or does the image seem to burst forward from the niche?

How close do you want to get? Why?

A Note on Nonobjective Sculpture. Until the twentieth century, sculpture used traditional materials—chiefly stone, wood, and clay—and was representational, imitating human beings or animals by means of masses of material. Sometimes the masses were created by cutting away (as in stone and wooden sculpture), sometimes they were created by adding on (as in clay sculpture, which then might serve as a model for a work cast in bronze), but in both cases the end result was a representation.

Twentieth-century sculpture, however, is of a different sort. For one thing, it is often made out of industrial products—plexiglass, celluloid, cardboard, brushed aluminum, galvanized steel, wire, and so forth—rather than made out of traditional materials, notably wood, stone, clay, and bronze. Second, instead of representing human beings or animals or perhaps ideals such as peace or war or death (ideals that in the past were often represented allegorically through images of figures), much twentieth-century sculpture is concerned with creating spaces. Instead of cutting away (carving) or building up (modeling) material to create representational masses, the sculptors join material (**assemblage**) to explore spaces or movement in space. Unlike traditional sculpture, which is usually mounted on a pedestal, announcing that it is a work of art, something to be contemplated as a thing apart from us, the more recent works we are now talking about may rest directly on the floor or ground, as part of the environment in which we move, or they may project from a wall or be suspended by a wire.

In a moment we will look at a work using nontraditional materials, but first let's consider a bit further this matter of nonrepresentational sculpture. Think of a traditional war memorial—for instance, a statue of a local general in a park, or the Iwo Jima Monument representing marines raising an American flag—and then compare such a work with Maya Lin's *Vietnam Veterans Memorial,* dedicated in 1982 (see page 56). Lin's pair of 200-foot granite walls join to make a wide V, embracing a gently sloping plot of ground. On the walls, which rise from ground level to a

Maya Lin, *Vietnam Veterans Memorial*, 1980–82. The Mall, Washington, D.C. Black granite. Each wall 10'1" × 246'9". (Corbis/Duane Preble)

height of about 10 feet at the vertex, are inscribed the names of the 57,939 Americans who died in the Vietnam War. (As visitors descend the slope to approach the wall with the names of the dead—a sort of descent into the grave—they experience a powerful sense of mortality.) Because this monument did not seem in any evident way to memorialize the heroism of those who died in the war, it stirred a great deal of controversy, and finally, as a concession to veterans groups, the Federal Fine Arts Commission came up with a compromise: A bronze sculpture of three larger-than-life armed soldiers (done by Frederick Hart) was placed nearby, thereby celebrating wartime heroism in a traditional way.

Although Lin's environmental sculpture has been interpreted as representational—the V-shaped walls have been seen as representing the chevron of the foot soldier, or as the antiwar sign of a V made with the fingers—clearly these interpretations are far-fetched. The memorial is not an object representing anything, nor is it an object that (set aside from the real world and showing the touch of the artist's hand) is meant to be looked at as a work of art, in the way that we look at a sculpture on a pedestal or at a picture in a frame. Rather, it is a *site*, a place for reflection. *Vietnam Veterans Memorial* belongs to a broad class of sculptures called *primary forms*. These are massive constructions, often designed in accordance with mathematical equations and often made by industrial fabricators. Robert Smithson's *Spiral Jetty* (see page 3 and the back cover) is another example, though it is made of earth and rocks rather than of industrial materials. The spiral, an archetypal form found in all cultures, is sometimes interpreted as suggesting an inward journey (the discovery of the self or the return to one's origins) or an outward journey into the cosmos.

Let's look at one other nonrepresentational work, Eva Hesse's *Hang Up* (1966), shown on page 58. Hesse, who died of a brain tumor in 1970 at the age of thirty-four, began as a painter but then turned to sculpture, and it is for her work as a sculptor that she is most highly regarded. Her materials were not those of traditional sculpture; Hesse used string, balloons, wire, latex-coated cloth, and other "non-art" materials to create works that (in her words) seem "silly" and "absurd." Only occasionally did Hesse create the sense of mass and sturdiness common in traditional sculpture; usually, as in *Hang Up*, she creates a sense that fragile things have been put together, assembled only temporarily. In *Hang Up*, a wooden frame is wrapped with bedsheets, and a half-inch metal tube, wrapped with cord, sweeps out (or straggles out) from the upper left and into the viewer's space, and then returns to the frame at the lower right. The whole, painted in varying shades of gray, has an ethereal look.

Taking a cue from Hesse, who in an interview with Cindy Nemser in *Artforum* (May 1970) said that she tried "to find the most absurd opposites or extreme opposites" and that she wanted to "take order versus chaos, stringy versus mass, huge versus small," we can see an evident opposition in the rigid, rectangular frame and the sprawling wire. There are also oppositions between the hard frame and its cloth wrapping or bandaging, and between the metal tubing and its cord wrapping. Further, there is an opposition or contradiction in a frame that hangs on a wall but that contains no picture. In fact, a viewer at first wonders if the frame *does* contain a panel painted the same color as the wall, and so the mind is stimulated by thoughts of illusion and reality. And although the work does not obviously represent any form found in the real world, the bandaging, and perhaps our knowledge of Hesse's illness, may put us in mind

Eva Hesse, American, b. Germany, 1936–1970, *Hang-Up*, acrylic
on cord and cloth, wood and steel, 1966. 182.9 × 213.4 × 198.1
cm. Through prior gifts of Arthur Keating and Mr. and Mrs.
Edward Morris, 1988.130. Photograph © 1999, The Art Institute
of Chicago. All rights reserved. © The Estate of Eva Hesse.

of the world of hospitals, of bodies in pain. (The materials that Hesse
commonly used, such as latex and fiberglass, often suggest the feel and
color of flesh.) In *Hang Up*, the tube, connected at each end to opposite
extremes of the swathed frame, may suggest a life-support system.

The title, too, provides a clue; *Hang Up* literally hangs on a wall, but
the title glances also at psychological difficulties—anyone's, but especially those of the artist, who was experiencing a difficult marriage and
who was trying to create a new form of sculpture. If we go back to the
idea of oppositions, we can say that the work itself has a hang-up: It

seems as though it wants to be a painting (the frame), but the painting never materialized and now the work is a sculpture.

In looking at nonobjective sculpture, consider the following:

- the scale (e.g., is it massive, like Lin's *Vietnam Veterans Memorial,* or on a more domestic scale, like *Hang Up?*)
- the effects of the materials (e.g., soft or hard, bright or dull?)
- the relationships between the parts (e.g., is the emphasis on masses or on planes, on closed volumes or on open assembly? If the work is an assembly, are light materials lightly put together, or are massive materials industrially joined?)
- the site (e.g., if the work is in a museum, does it hang on a wall or does it rest without a pedestal on the floor? If it is in the open, what does the site do to the work, and what does the work do to the site?)
- the title (e.g., is the title playful? enigmatic? significant?)
- comments by the sculptor, such as may be found in the four collections of statements by artists, mentioned on page 47.

A Cautionary Word about Photographs of Sculpture. Photographs of works of sculpture are an enormous aid; we can see details of a work that, even if we were in its presence, might be invisible because the work is high above us on a wall or because it is shrouded in darkness. The sculptural programs on medieval buildings, barely visible *in situ,* can be analyzed (e.g., for their iconography and their style) by means of photographs. But keep in mind the following points:

- Because a photograph is two-dimensional, it gives little sense of a sculpture in the round.
- A photograph may omit or falsify color, and it may obliterate distinctive textures.
- The photographer's lighting may produce dramatic highlights or contrasts, or it may (by even lighting) eliminate highlights that would normally be evident. Further, a bust (say, a Greek head in a museum) when photographed against a dark background may seem to float mysteriously, creating an effect very different from the rather dry image of the same bust photographed, with its mount visible, against a light gray background.
- A photograph of a work even in its original context (to say nothing of a photograph of a work in a museum) may decontextualize the work, such as by not taking account of the

angle from which the work was supposed to be seen. The first viewers of Michelangelo's *Moses* had to look upward to see the image, but almost all photographs in books show it taken straight-on. Similarly, a photo of Daniel Chester French's *Lincoln* can convey almost nothing of the experience of encountering the work as one mounts the steps of the Lincoln Memorial.

- Generally, photographs do not afford a sense of scale; for example, one may see Michelangelo's *David* (about 13 feet tall) as no bigger than a toy soldier, unless, as in the unusual photograph on page 28, human viewers are included.

Architecture

You may recall from Chapter 1 (page 8) Auden's comment that a critic can "throw light upon the relation of art to life, to science, economics, religion, etc." Works of architecture, since they are created for use, especially can be considered in the context of the society that produced them. As the architect Louis Sullivan (1856–1924) said, "Once you learn to look upon architecture not merely as an art, more or less well or badly done, but as a social manifestation, the critical eve becomes clairvoyant, and obscure, unnoted phenomena become illumined."

The Roman architect Vitruvius suggested that buildings can be judged according to their

- *utilitas* (function, fitness for their purpose)
- *firmitas* (firmness, structural soundness)
- *venustas* (beauty, design)

Utilitas gets us thinking about how suitable (convenient, usable) the building is for its purposes. Does the building work, as (say) a bank, a church, a residence, a school? *Firmitas* gets us thinking about the structure and the durability of the materials in a given climate. *Venustas* gets us thinking about the degree to which it offers delight. Much (though not all) of what follows is an amplification of these three topics.

What did the client want? What was or is the **purpose** of the building? For instance, does it provide a residence for a ruler, a place of worship, or a place for legislators to assemble? Was this also its original purpose? If not, what was the building originally used for? Consider Le Corbusier's maxim, "A house is a machine for living in" (*Une maison est une machine-à-habiter*). All architecture is designed to help us to live—even a tomb is designed to help the living to cope with death, perhaps by assuring them that the deceased lives in memory. Churches, museums, theaters, banks, zoos, schools, garages, residences, all are designed to facilitate the business of living.

Does the building appear today as it did when constructed? Has it been added to, renovated, restored, or otherwise changed in form? **What does the building say?** "All architecture," wrote John Ruskin, "proposes an effect on the human mind, not merely a service to the human frame." One can distinguish between function as housing and function as getting across the patron's message. A nineteenth-century bank said, by means of its bulk, its bronze doors, and its barred windows, that your money was safe; it also said, since it had the façade of a Greek temple, that money was holy. A modern bank, full of glass and color, is more likely to say that money is fun. Some older libraries look like Romanesque churches, and the massive J. Edgar Hoover FBI building in Washington, D.C., with its masses of precast concrete, looks like a fortress, uninviting, menacing, impregnable—the very image of the FBI.

What, then, are the architectural traditions behind the building that contribute to the building's expressiveness? The Boston City Hall (see page 64), for all its modernity and (in its lower part) energetic vitality, is tied together in its upper stories by forceful bands of windows, similar in their effect to a classical building with columns. (Classical façades, with columns, pediments, and arches, are by no means out of date. One can see versions of them, often slightly jazzed up, as the entrances to malls and high-style retail shops that wish to suggest that their goods are both timeless and in excellent taste.)

Here is how Eugene J. Johnson sees (or hears) Mies van der Rohe's Seagram Building (1954–57):

> Austere, impersonal, and lavishly bronzed, it sums up the power, personality, and wealth of the modern corporation, whose public philanthropy is symbolized by the piazza in front, with its paired fountains— private land donated to the urban populace. If the piazza and twin fountains call to mind the Palazzo Farnese in Rome, so be it, particularly when one looks out from the Seagram lobby across an open space to the Renaissance-revival façade of McKim, Mead and White's Racquet Club which quotes the garden façade of Palazzo Farnese! Mies set up a brilliant conversation between two classicizing buildings, bringing the nineteenth and twentieth centuries together without compromise on either part. Mies was in many ways *the* great classicist of this century. One might say that one of his major successes lay in fusing the

principles of the great classical tradition of Western architecture with the raw technology of the modern age.

—"United States of America," in *International Handbook of Contemporary Developments in Architecture,* ed. Warren Sanderson (1981), 506

Notice how Johnson fuses *description* (e.g., "lavishly bronzed") with interpretive *analysis* (he sees in the bronze the suggestion of corporate power). Notice, too, how he connects the building with history (the debt of the piazza and twin fountains to the Palazzo Farnese), and how he connects it with its site (the "conversation" with a nearby building).

Like other buildings, **museums and the exhibition spaces** within museums make statements. This is true whether the museum resembles a Greek temple and is entered only after a heavenward ascent up a great flight of steps, or whether it is a Renaissance palace or a modern imitation of one, or whether it is insistently high-tech. In thinking about an exhibition it is usually worth asking oneself what sorts of statements the museum and the exhibition space make. Think about why the material is being presented (consider the difference between titling an exhibition "Festival of Indonesia" and "100 Masterpieces from Indonesia") and why it is presented in this particular way. For instance, is the material presented with abundant verbal information on the walls, perhaps thereby emphasizing the cultural context, or with minimal labels and lots of empty space on the walls, perhaps thereby emphasizing the formal properties and the aesthetic values of each work. (For an examination of many museums built in recent decades, see Victoria Newhouse, *Toward the New Museum,* 1998. Among the topics Newhouse considers are "The Museum as Sacred Space," "The Museum as Entertainment," and "The Museum as Environmental Art.")

Do the forms and materials of the building relate to its neighborhood? What does the building contribute to the **site?** What does the site contribute to the building? Does the building contrast with the site or complement it? How big is the building in relation to the neighborhood, and in relation to human beings; that is, what is the **scale?** The Cambridge City Hall (1889), atop a slope above the street, crowns the site and announces—especially because it is in a Romanesque style—that it is a bastion of order, even of piety, giving moral significance to the neighborhood below. The Boston City Hall (1968), its lower part in brick, rises out of a brick plaza—the plaza flows into spaces between the concrete pillars that support the building—and seems to invite the crowds from the

Cambridge City Hall, 1889. (Photograph by G. M. Cushing. In *Survey of Architectural History in Cambridge, Report Two: Mid-Cambridge* [Cambridge Historical Commission, 1967], 44)

neighboring shops, outdoor cafés, and marketplace to come in for a look at government of the people by the people, and yet at the same time the building announces its importance.

How do you **approach** the building, and how do you enter it?

Does **form** follow **function?** For better or for worse? For example, does the function of a room determine its shape? Are there rooms with geometric shapes irrelevant to their purposes? Louis Sullivan, in rejecting the nineteenth-century emphasis on architecture as a matter of style (in its most extreme form, a matter of superficial decoration), said, "Form ever follows function." Sullivan's comment, with its emphasis on structural integrity—the appearance of a building was to reveal the nature of

Boston City Hall.
City Hall Plaza
Center, 1961–68.
(Photograph by
Cervin Robinson. In
Architecture Boston
[Boston Society of
Architects, 1976], 6)

its construction as well as the nature of its functions or uses—became the slogan for many architects working in the mid-twentieth century. But there were other views. For instance, Philip Johnson countered Sullivan with "forms always follows forms and not function." In looking at a building, ask yourself if the form serves a **symbolic function;** recall the "lavishly bronzed" Seagram building, which Eugene J. Johnson sees as summing up "the power, personality, and wealth of the modern corporation." Equally symbolic or expressive is Eero Saarinen's Trans World Airlines Terminal at Kennedy Airport, where the two outstretched wings suggest flight. Consider, too, Frank Gehry's Solomon R. Guggenheim Museum (1997) in Bilbao, built by the side of a river in a Spanish port city whose economy was based on shipbuilding. Viewed from across the river, the titanium-sheathed design evokes a ship (opposite), expressing such ideas as the history of the city, grandeur, and elegance.

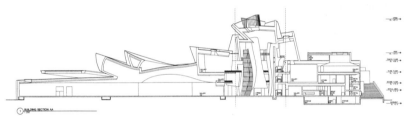

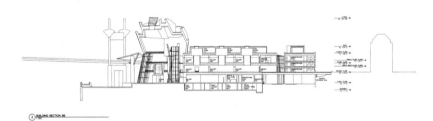

(*above*) Frank Gehry, Solomon R. Guggenheim Museum, 1997. Bilbao. The titanium-sheathed design evokes a ship, an apt form for a building in a city whose economy was largely based on shipbuilding and steel. (Guggenheim Museum Bilbao, exterior view. photograph by David Heald © The Solomon R. Guggenheim Foundation, New York.) (*below*) Detail of architect's rendering (Courtesy Frank Gehry)

What **materials** are used? How do the materials contribute to the building's purpose and statement? Take, for instance, the materials of some college and university buildings. Adobe works well at the University of New Mexico, but would it be right for the Air Force Academy in Colorado? (The Academy uses different materials—notably aluminum, steel, and glass.) Brick buildings in Collegiate Georgian suggest a connection with the nation's earliest colleges; stone buildings in Collegiate Gothic (pointed arches and narrow windows) are supposed to suggest a preindustrial world of spirituality and scholarship.

The idea that *marble*—the material of the Parthenon—confers prestige dies hard: The Sam Rayburn House Office Building in Washington, D.C., is clad in marble veneer (costing many millions of dollars) because marble is thought to suggest dignity and permanence—though in fact marble is a rather soft stone. Marble is highly versatile: White or black marble, common in expensive jewelry shops, can seem sleek or aloof; pink or creamy marble, in a boutique with goods for women, can seem soft and warm. Each building material has associations, or at least potential associations. *Brick,* for instance, often suggests warmth or unpretentiousness and handcraftsmanship. *Wood,* like marble, is amazingly versatile. In its rough-bark state is suggests the great outdoors; trimmed and painted it can be the clapboard and shingles of an earlier America; smooth and sleek and unpainted it can suggest Japanese elegance. *Glass* can be transparent, translucent, or even opaque as in I. M. Pei's John Hancock Building in Boston. The exterior walls of the Hancock building reflect the sky and clouds, thereby animating and softening the building which towers above its neighbors.

Do the exterior walls seem hard or soft, cold or warm? Is the sense of hardness or coldness appropriate? (Don't simply assume that metal must look cold. A metal surface that reflects images can be bright, lively, and playful. Curves and arches of metal can seem warm and "soft.") Does the material in the interior have affinities with that of the exterior? If so, for better or for worse? (Our experience of an interior brick wall may be very different from our experience of an exterior brick wall.)

Does **the exterior** stand as a massive sculpture, masking the spaces and the activities within, or does it express them? The exterior of the Boston City Hall (see page 64) emphatically announces that the building harbors a variety of activities; in addition to containing offices, it contains conference rooms, meeting halls, an exhibition gallery, a reference library, and other facilities. Are the spaces continuous? Or are they static, each volume capped with its own roof?

What is the function of **ornament,** or of any **architectural statuary** in or near the building? To embellish a surface? To conceal the joins of a surface? To emphasize importance? (The east end of a Christian church, where the altar is, sometimes is more elaborately decorated than the rest of the building.)

Does the interior arrangement of spaces say something—for example, is the mayor's office in the city hall on the top floor, indicating that he or she is above such humdrum activities as dog licensing, which is on the first floor?

In a given room, what is the function of the walls? To support the ceiling or the roof? To afford privacy and protection from weather? To provide a surface on which to hang shelves, blackboards, pictures? If glass, to provide a view in—or out?

What is the effect of the floor (wood, tile, brick, marble, carpet)? Notice especially the effect if you move from a room floored with one material (say, wood) to another (say, carpet).

Is the building inviting? The architect Louis Kahn said, "A building should be a . . . stable and *harboring* thing. If you can now [with structural steel] put columns as much as 100 feet apart you may lose more than you gain because the sense of enclosed space disappears." Is the public invited? The Cambridge City Hall has one public entrance, approached by a flight of steps; the Boston City Hall, its lower floor paved with the brick of the plaza, has many entrances, at ground level. What are the implications in this difference?

"There is no excellent Beauty that hath not some strangeness in the proportion," said Francis Bacon in the early seventeenth century. Does the building evoke and then satisfy a sense of curiosity?

What is the role of **color?** To clarify form? To give sensuous pleasure? To symbolize meaning? (Something has already been said, on page 66, about the effects of the colors of marble.) Much of the criticism of the *Vietnam Veterans Memorial* centered on the color of the stone walls. One critic, asserting that "black is the universal color of shame [and] sorrow," called for a white memorial.

What part does the changing **daylight** play in the appearance of the exterior of the building? Does the interior make interesting use of natural light? And how light is the interior? (The Lincoln Memorial, open only at the front, is somber within, but the Jefferson Memorial, admitting light from all sides, is airy and suggestive of Jefferson's rational—sunny, we might say—view of life. Similarly, the light in a place of worship differs from the light in a classroom.)

As the preceding discussion suggests, architectural criticism usually
is based on three topics:

- the building or monument as an envelope (its purpose, structural
  system, materials, sources of design, history, design [articulation
  of the façade, including the arrangement of the windows and
  doors, ornamentation, color])
- the interior (hierarchy of spaces, flow of traffic, connection with
  the exterior)
- the site (relationship of the building to the environment)

A fourth topic is

- the architect's philosophy and the place of the building in the
  history of the architect's work

If you are writing about the first or second of these topics—the building
as an envelope or the enclosed spaces through which one moves—you
may have only your eyes and legs to guide you when you study the building of your choice, say, a local church or a college building. But if the
building is of considerable historical or aesthetic interest, you may be
able to find a published *plan* (a scale drawing of a floor, showing the
arrangement of the spaces) or an *elevation* (a scale drawing of an external
or internal wall). Plans and elevations, often available in printed sources,
are immensely useful as aids in understanding buildings. For a helpful introductory discussion of architecture that makes excellent use of plans
and elevations, see Simon Unwin, *Analysing Architecture* (1997).

A few words about the organization of an essay on a building may be
useful. Much will depend on your purpose, and on the building, but consider the possibility of using one of these three methods of organization:

1. You might discuss, in this order, *utilitas, firmitas,* and
   venustas—that is, function, structure, and design (see page 60).
2. You might begin with a view of the building as seen at a
   considerable distance, then at a closer view, and then go on to
   work from the ground up, since the building supports itself this
   way.
3. You might take, in this order, these topics:

- the materials (smooth or rough, light or dark, and so on)
- the general form, perceived as one walks around the building (e.g., are the shapes square, rectangular, or circular, or what? Are they simple or complex?)
- the façades, beginning with the entrance (e.g., is the entrance dominant or recessive? How is each façade organized? Is there variety or regularity among the parts?)
- the relation to the site, including materials and scale

Photography

A *photograph* is literally an image "written by light." Although most people today think of a photograph as a flat work on paper produced from a *negative* (in which tones or colors are the opposite from what we normally see) made in a camera, none of these qualities is necessary.° Photographs can be made without cameras (by exposing a light-sensitive material directly to light), can be generated without negatives (color slides and Polaroid prints are familiar examples of what are known as *direct positives*), and can be pieces of fabric, leather, glass, metal, or even a still image on a computer screen. All that you need to have a photograph is a substance that changes its color or tone under the influence of light and that can subsequently be made insensitive to light so that it can be viewed.

Inherent in this requirement that the image be created by light is the fundamental distinction between photography and prior ways of making pictures. Instead of having the hand drag oil paint across a stretched piece of linen or incise lines into a metal plate to generate an engraving, nature itself, with the help of some manufactured lenses and chemically coated sheets of glass, paper, or film, became the artist. Early photographic viewers marveled at the seemingly limitless details that magically appeared on the metal plate used for one of the first processes, the *daguerreotype*. Thus was born the idea that the camera cannot lie and that the image it produced depended on chemistry and optics, not on human skills.

°This discussion of photography is by Elizabeth Anne McCauley (University of Massachusetts-Boston), author of numerous studies of the history of photography, including *Industrial Madness: Commercial Photography in Paris, 1848–1871* (New Haven, Conn.: Yale University Press, 1994).

At the same time, photographers and critics who were familiar with the craft realized that there was a huge gap between what the eye saw and the finished photograph. Human beings have two eyes that are constantly moving to track forms across and into space; they perceive through time, not in fixed units; their angle of vision (the horizontal span perceived when holding the eye immobile) is not necessarily that of a lens; their eyes adjust rapidly to read objects both in bright sun and deep shadows. As Joel Snyder and Neil Allen observed in *Critical Inquiry* (Autumn 1975), "a photograph shows us 'what we would have seen' at a certain moment in time, *from* a certain vantage point *if* we kept our head immobile *and* closed one eye *and if* we saw with the equivalent of a 150-mm or 24-mm lens *and if* we saw things in Agfacolor or in Tri-X developed in D-76 and printed on Kodabromide #3 paper" (page 152). And, they point out, if the eye and the camera saw the world in the same way, then the world would look the way it does in photographs.

Photographers also intervene in every step of the photographic process. They pose sitters; select the time of day or artificial light source; pick the camera, film, exposure time, and amount of light allowed to enter the camera (the lens *aperture*); position the camera at a certain height or distance from the subject; focus on a given plane or area; develop the film to bring out certain features; choose the printing paper and manipulate the print during enlargement and development; and so forth. By controlling the world in front of the camera, the environment within the camera, and the various procedures after the light-sensitive material's initial exposure to light, the photographer may attempt to communicate a personal interpretation or vision of the world. If he or she succeeds, we may begin to talk about a photographic "style" that may be perceptible in many images made over several years.

Since its invention in the early nineteenth century, photography has taken over many of the functions of the traditional pictorial arts while satisfying new functions, such as selling products or recording events as they actually happen. Because early photography seemed close to observed reality, it was rapidly used any time factual truth was required. Photographs have been taken of ancient and modern architecture, engineering feats, criminals' faces, biological specimens, astronomical phenomena, artworks, slum conditions, and just about anything that needed to be classified, studied, regulated, or commemorated. Because the validity of these images depended on their acceptance as truth, their creators often downplayed their role in constructing the photographs. The photojournalist Robert Capa, talking about his famous views of the Spanish Civil

War, said, "No tricks are necessary to take pictures in Spain. . . . The pictures are there, and you just take them. The truth is the best picture, the best propaganda."

While statements such as Capa's reveal what photographers once wanted the public to believe about their images, we now no longer accept photographs, even so-called documentary ones, as unmanipulated truth. All photographs are representations, in that they tell us as much about the photographer, the technology used to produce the image, and their intended uses as they tell us about the events or things depicted. In some news photographs, we now know that the event shown was in part staged. For example, the British photographer Roger Fenton, who was one of the first cameramen to record a war, moved the cannonballs that litter the blasted landscape in his famous *Valley of the Shadow of Death,* a photograph taken in 1855 during the Crimean War. Does this make a difference when we look at the picture as evidence of how a battleground appeared? Perhaps not. But we should be careful about ever assuming that from photographic evidence we can always draw valid conclusions about the lives of people, the historical meaning of events, and the possible actions that we should take.

Let's look now at a photograph (shown on page 72) by Dorothea Lange, an American photographer who made her reputation with photographs of migrant farmers in California during the Depression that began in 1929. Lange's *Migrant Mother, Nipomo, California* (1936) is probably the best-known image of the period. A student made the following entry in a journal in which he regularly jotted down his thoughts about the material in an art course he was taking. (The student was given no information about the photograph other than its title and date.)

> This woman seems to be thinking. In a way, the picture reminds me of a statue called The Thinker, of a seated man who is bent over, with his chin resting on his hand. But I wouldn't say that this photograph is really so much about thinking as it is about other things. I'd say that it is about several other things. First (but not really in any particular order), fear. The children must be afraid, since they have turned to their mother. Second, the picture is about love. The children press against their mother, sure of her love. The mother does not actually show her love--for instance, by kissing them, or even hugging them--but you feel she loves them. Third, the picture is about hopelessness. The mother doesn't seem to be able to offer any comfort. Probably they have very little food; maybe they are homeless. I'd say the picture is also about

Dorothea Lange, *Migrant Mother, Nipomo, California.* Nipomo, California, USA, North America. (1936) Gelatin-silver print, 12 1/2 × 9 7/8" (31.9 × 25.2 cm). The Museum of Modern Art, New York. Purchase. Copy Print. © 1999 The Museum of Modern Art, New York.

courage. Although the picture seems to me to show hopelessness, I also think the mother, even though she does not know how she will be able to help her children, shows great strength in her face. She also has a lot of dignity. She hasn't broken down in front of the children; she is going to do her best to get through the day and the next day and the next.

Another student wrote:

Is this picture sentimental? I remember from American Lit that good literature is not sentimental. (When we discussed the word, we concluded that "sentimental" meant "sickeningly sweet.") Some people might think that Lange's picture, showing a mother and two little children, is sentimental, but I don't think so. Although the children must be upset, and maybe they even are crying, the mother seems to be very strong. I feel that with a mother like this, the children are in very good hands. She is not "sickeningly sweet." She may be almost overcome with despair, but she doesn't seem to ask us to pity her.

A third student wrote:

> Why does this picture bother me? It's like those pictures of the homeless in the newspapers and on TV. A photographer sees some man sleeping in a cardboard box, or a woman with shopping bags sitting in a doorway, and he takes their picture. I suppose the photographer could say that he is calling the public's attention to "the plight of the homeless," but I'm not convinced that he's doing anything more than making money by selling photographs. Homeless people have almost no privacy, and then some photographer comes along and invades even their doorways and cardboard houses. Sometimes the people are sleeping, or even if they are awake they may be in so much despair that they don't bother to tell the photographer to get lost. Or they may be mentally ill and don't know what's happening. In the case of this picture, the woman is not asleep, but she seems so preoccupied that she isn't aware of the photographer. Maybe she has just been told there is no work for her, or maybe she has been told she can't stay if she keeps the children. Should the photographer have intruded on this woman's sorrow? This picture may be art, but it bothers me.

All of these entries are thoughtful, interesting, and helpful—material that can be the basis of an essay—though of course even taken together they do not provide the last word. Far from being a neutral document, this photograph encapsulates Lange's sympathy for displaced migrant workers and her belief that they need government assistance. As comparisons between this justly famous image and other prints Lange took at the same time reveal, only this photograph is symmetrical and contrasts the strained face of the mother with the two bedraggled but anonymous children and the almost hidden baby in her lap. Although the clothing, place, and people are real, in the sense that Lange found them in this condition, she undoubtedly encouraged this pose, which echoes traditional representations of the Madonna and child. The tow-headed but grimy children become all children; the mother, seeming to look searchingly off to her right while raising her worn hand tentatively to her chin, could be any American mother worried about her family. The most effective social document, as Lange herself would readily admit, is not necessarily the spontaneous snapshot taken without the photographer's intervention in the scene. It is the careful combination of subject matter and composition that grabs our attention and holds it.

Some photographs are more obvious in the ways that they reveal the point of view of the person behind the camera and the manipulation of observed reality. Many of these images are intended to be sold and appreciated as aesthetic documents. In other words, the formal properties of the image—its composition, shades of tone or color, quality of light, use of blurs and grain—become as important as or more important than the depicted subject in inspiring feelings or ideas in the viewer. Edward Weston's *Pepper #30* (1930), placed against a mysterious black background and softly lit from the side, makes us think of the ways that the ever-changing curves and bulges of common vegetables repeat the sensual, intertwined limbs of the human body. By moving close to the isolated object and carefully controlling the lighting, Weston assumes an aesthetic rather than a botanical approach. The picture would not be very successful as an illustration for a book on the various species of peppers and their reproductive structures; it does not present the kind of visual information that we have grown to expect from such illustrations.

In other cases, the distinctions between the artistic photograph and the documentary photograph are less clear from internal evidence. Many nineteenth-century photographs that were originally conceived as records of buildings or machinery have struck recent viewers as beautiful and expressive of a personal vision. The turn-of-the-century French photographer Eugène Atget earned his living selling standard-sized prints of eighteenth-century Parisian architecture and rapidly disappearing street vendors to antiquarians, libraries, cartoonists, and illustrators. By the time of Atget's death in 1927, avant-garde artists were struck by the uncanny stillness and radiant light of many of his photographs and began collecting them as the work of a naive genius. As tastes have changed and the intended functions of photographic images have been forgotten or ignored, our understanding of photography has also shifted from the functional or documentary to the aesthetic. In writing about photographs, you should be aware of the ways that the passage of time and the changes in context transform what photographs mean. We do not respond to a framed photograph on the walls of a museum in the same way that we respond to that photograph reproduced in *Life* magazine.

At various points in the history of the medium, photographers have pointedly taken past or contemporary paintings or drawings as models for artistic photographs. They introduced Rembrandtesque lighting into portraits or contrived figure compositions imitating Raphael or Millet. In the late nineteenth century, to distinguish their works from the flood of easily made snapshots of family outings, self-proclaimed art photographers

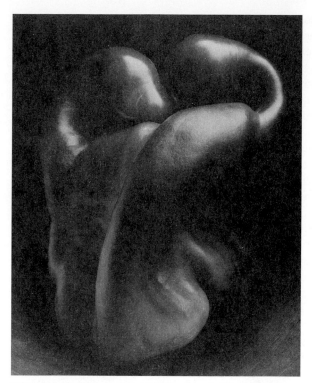

Edward Weston, *Pepper #30, 1930.* Photograph. Weston used long
time exposures for his peppers, ranging from six minutes to fifty. In
an entry dated 24 April 1930 he repeated in his journal, which he
called *Daybook,* a statement that he had written for a museum:
"Clouds, torsos, shells, peppers, trees, rocks, smoke stacks, are but
interdependent, interrelated parts of a whole, which is life.—Life
rhythms felt in no matter what, become symbols of the whole."
And on 8 August 1930 he wrote, "It is a classic, completely
satisfying—a pepper—but more than a pepper: abstract, in that it
is completely outside subject matter . . . this new pepper takes one
beyond the world we know in the conscious mind." (The Lane
Collection, courtesy, Museum of Fine Arts, Boston)

used soft-focus and extensive hand manipulation of negatives and prints
to make photographs that looked like charcoal drawings. These *pictorial-
ists,* whose goal was to elevate photography to the rank of serious art by
imitating the look of that art, were condemned in the early twentieth century by such photographers as Alfred Stieglitz (1864–1946), who argued

that the best photographs were those that were honest to the unique properties of camera vision. (There was, however, little recognition that these properties were themselves changing as equipment and processes changed.)

Whereas modern photographers have rarely tried to legitimize their practice by literally copying famous paintings, they have often reintroduced manual manipulations and elaborate staging. A **manipulated photograph** is one in which either the negative or the positive (i.e., the print) has been altered by hand or, more recently, by computer. Negatives may be scratched, drawn on, or pieced together; positive prints may be hand-tinted or have their emulsions (the coating on the paper that contains the light-sensitive material) smeared with a brush. Computer-aided processes are now often used to manipulate images. For instance, images from old magazines can be put into digital form and combined and colorized on a computer, allowing the artist, in David Hockney's words, "to work on photographs the way a draftsman might" (*Aperture*, 1992). Yasumasa Morimura, one of the most ingenious makers of manipulated pictures, uses computer technology to create composite photographs of famous paintings with images of himself—usually costumed in the appropriate period or semi-nude. Thus, for his *Portrait (Nine Faces)*, he used a computer to blend parts of his own face into the nine faces (including the face of the corpse) of Rembrandt's *The Anatomy Lesson*. A **fabricated photograph** is one in which the subject has been constructed or staged to be photographed. In a certain sense, all still lifes are fabricated photographs. More recently, photographers coming out of art school backgrounds have choreographed events or crafted environments to challenge our belief in photographic truth or question conventions of representation. For example, William Wegman made a jungle out of a couple of plants and turned his dog into an elephant by means of a long sock put on to the dog's snout.

When we turn to analyze a photograph, many of the questions on drawing and painting—for example, those on composition and color—also apply. But because of the different ways that photographs are produced, their peculiar relationship to the physical world as it existed at some point in the past, and their multiple functions, we need to expand our questions to reflect these special concerns.

Our first concern in analyzing a photograph is **identification of the work** (much of this information is often provided by a wall label or photo caption). Who took the photograph? Was it an individual or a photographic firm (usually identified on the negative or in a stamp or label)?

Was the positive print produced by the same individual (or firm) who exposed the negative? If the photographer is unknown, can the photograph be assigned to a country or region of production? What is the **title** of the work? Was the photograph given a title by its creator(s) in the form of an inscription on the negative or positive? Or was the title added by later viewers? What does the title tell you about the intended function of the photograph? Did the photographer identify the **date** of the work? When was the exposure made? When was the positive print made?

What type of **photographic process** was used to produce the negative? The positive? On what sort of paper was the positive printed? Are these processes typical for the period? Why do you think the photographer chose these particular processes?

Analysis of a photographic work also involves looking at its **physical properties.** What are the dimensions of the photograph? Are these typical for the techniques chosen? What effect does the **size** of the work have on the viewer's analysis of it? Is the photograph printed on paper or on some other type of material, such as metal or silk? If the image is on paper, is the paper matte, glossy, or somewhere in between? What is the color of the print? Has it been hand-tinted or retouched? How do the physical properties of the print influence the viewer's reaction? Is the print damaged, torn, abraded, faded? Has the paper been trimmed or cropped? If the image is on metal, has there been corrosion, tarnishing, or other damage that alters your appreciation of the image? Are there signs in the print of damage to the negative?

Briefly describe the **subject** of the photograph. Is it a traditional subject—a landscape, a still life, a portrait, a genre, an allegorical or historical scene, or a documentary? Do the figures or objects in the picture seem arranged by the photographer or caught "as they were"? Are props included? To what extent did the photographer fabricate or create the image by physically constructing, arranging, or interacting with some or all of its components? How does this affect the viewer's response?

In addition, the many **formal properties** of a photograph are relevant to an analysis. If the work is considered as a **two-dimensional** composition, how is the subject represented? Which are the most important forms and where are they located on the picture's plane? Is the composition balanced or unbalanced? How does your eye read the photograph? What did the photographer leave out of the frame? What happens along the edges of the photograph? How obvious and important is the two-dimensional composition?

If the work is considered as a **three-dimensional** composition, where is the main activity taking place—in the foreground, the midground, the background, or a combination of these areas? How did the photographer define (or not define) the three-dimensional space? How can you describe the space—as shallow or deep, static or dynamic, claustrophobic or open, rational or irrational? How important is the three-dimensional composition?

Look for the photographer's choice of **vantage point** and **angle of vision.** How near or far does the main subject appear? Does the photograph draw the viewer's attention to where the photographer was located? How does the position from which the picture was taken contribute to the mood or content of the image? What is the angle of vision? How does it compare with "normal" vision? Is there lens distortion? Why do you think the photographer chose the lens he or she did?

Examine the **detail** and **focus** of the work. Can you characterize the overall focus? Where are the areas of sharp focus? Soft focus? What is the **depth of field** (i.e., the minimum and maximum distances from the camera that are in sharp focus)? How did the photographer use focus to convey meaning? Is the image detailed or grainy? How does the detail or grain contribute to or detract from the image?

What sort of **lighting** was used? Was the photograph taken out-of-doors or inside, in natural or artificial light (or in some combination of the two)? How did the season or time of day affect the lighting conditions? Where is the main light source? The secondary light source? Was a flash used? How can you characterize the lighting—as harsh, subtle, flat, dramatic, magical, or what? What effects do you think the photographer was trying to achieve through the use or control of light?

Consider also **contrast** and **tonal range.** Within the black and white print, what is the range of light and dark? Where are the darkest and lightest areas? Does the print have high contrast, with large differences in tone from light to dark, or does it have low contrast, with many shades of gray? What overall effect do contrast and tone create?

What **exposure time** did the photographer choose? How does the length of the exposure influence the image? Do any blurs or midaction motions signal the passage of time? How does the length of the exposure add or detract from the image?

Taking into consideration the preceding points, what do you think the photographer was trying to say in this image? What aspects of the subject did the photographer want to accentuate? What was the photographer's attitude toward the subject? What does the photograph convey to you today—about a place, a time, a person, an event, or a culture?

Video Art

Video art originated in 1965, when Nam June Paik (musician, sculptor, filmmaker) used the Sony Portapak, a portable videotape system, to record from a taxi the visit of Pope Paul VI to New York. He then showed the tapes at a New York café frequented by artists. The announcement for the screening said, "Some day artists will work with capacitors, resistors, and conductors just as they work today with brushes, violins, and junk." Video art is a medium, like oil painting or photography, not a style; the color can be intense or washed out, the images can move or be still, the recorded material can be spontaneous (as in television news coverage) or highly scripted (as in a sitcom), the work can celebrate popular culture or it can be critical of it. It uses the technology of commercial television but for purposes regarded as artistic rather than commercial.

The monitor showing the tape may itself be part of a larger work, in which case the whole can be considered a sculpture. Thus, Paik's *Electronic Superhighway* (1995), fifteen feet tall and thirty-two feet long, uses neon to represent a map of the United States, and within what might almost be called a neon drawing are set dozens of television monitors and laser disk images. Two other works by Paik: In *TV Buddha* (1982) the

Nam June Paik, *TV Buddha*, 1982. Video. (Courtesy of the Holly Solomon Gallery)

monitor records the live camera's image of the motionless head of a Bud-dha who appears to be contemplating his own image on the monitor; in *Video Fish* (1975), five tanks with live fish are in front of five monitors with video tapes of swimming fish. Thus, the tanks seem to be monitors and the monitors seem to be tanks, and the whole (like *TV Buddha*) stim-ulates the viewer to meditate on relationships between reality and repre-sentation, and, indeed, to meditate on the relationships between medita-tion and technology.

In writing about video art, consider

- the visual impact (e.g., the work as sculpture)
- the use of sound (music, talk, or noise usually is part of the work)
- the context (Is the work shown in a museum—and if so, in one of the galleries or in the cafeteria—or on the street, or in a café? If the work is shown on a screen in your home, how does this context relate to the video?)
- the political implications (much video art satirizes bourgeois interests)
- the connections with earlier art history (e.g., with documentary film, or with surrealism)

For statements by video artists, see *Video Art: An Anthology*, edited by Ira Schneider and Beryl Korot (1976). For an exhibition catalog with im-ages and useful essays, see *Video Art* (1975; no editor, but essays by David Antin and others).

Another Look at the Questions

As the preceding discussion has shown, there are many ways of helping yourself to see. In short, you can stimulate responses (and understand-ing) by asking yourself two basic questions:

- *What is this doing?* Why is this figure here and not there? Why is the work in bronze rather than in marble? Or put it this way: What is the artist up to?
- *Why do I have this response?* Why do I find this landscape op-pressive but that landscape inviting, this child sentimental but that child fascinating? That is, how did the artist manipulate the mate-rials in order to produce the strong feelings that I experience?

The first of these questions (*What is this doing?*) requires you to identify yourself with the artist, wondering, perhaps, why the artist chose one medium over another, whether pen is better than pencil for this drawing, or watercolor better than oil paint for this painting.

Sometimes artists tell us what they are up to. Van Gogh, for example, in a letter (11 August 1888) to his brother, helps us to understand why he put a blue background behind the portrait of a blond artist: "Behind the head instead of painting the ordinary wall of the mean room, I paint infinity, a plain background of the richest, intensest blue that I can contrive, and by the simple combination of the bright head against the rich blue background, I get a mysterious effect, like a star in the depths of an azure sky." But, of course, you cannot assume that the artist's stated intention has been fulfilled in the work itself.

The second question (*Why do I have this response?*) requires you to trust your feelings. If you are amused or repelled or unnerved or soothed, assume that your response is appropriate and follow it up—but not so rigidly that you exclude the possibility of other, even contradictory feelings. (The important complement to "Trust your feelings" is "Trust the work of art." The study of art ought to enlarge feelings, not merely confirm them.)

Almost any art history book that you come across will attempt to answer questions posed by the author. For example, in the introduction to *American Genre Painting: The Politics of Everyday Life* (1991), Elizabeth Johns writes:

> Two simple questions underscore my diagnosis: "Just whose 'everyday life' is depicted?" and "What is the relationship of the actors in this 'everyday life' to the viewers?"

The book contains her answers.

Indeed, most art historians ask the questions "What?" "Why?" and "Who?"—and offer answers.

FORMAL ANALYSIS

What Formal Analysis Is

It should be understood that the word *formal* in **formal analysis** is not used as the opposite of *informal,* as in a formal dinner or a formal dance. Rather, a formal analysis is an analysis of the form the artist produces; that is, an analysis of the work of art, which is made up of such things as line, shape, color, texture, mass, composition. These things give the stone or canvas its form, its expression, its content, its meaning. Rudolf Arn-heim's assertion that the curves in Michelangelo's *The Creation of Adam* convey "transmitted, life-giving energy" is a brief example. (See page 30.) Similarly, one might say that a pyramid resting on its base conveys stabil-ity, whereas an inverted pyramid—one resting on a point—conveys insta-bility or precariousness. Even if we grant that these forms may not uni-versally carry these meanings, we can perhaps agree that at least in our culture they do. That is, members of a given *interpretive community* per-ceive certain forms or lines or colors or whatever in a certain way.

Formal analysis assumes a work of art is (1) a constructed object (2) with a stable meaning (3) that can be ascertained by studying the rela-tionships between the elements of the work. If the elements "cohere," the work is "meaningful." That is, the work of art speaks directly to us, and we understand its language—we respond appropriately to its charac-teristics (shape, color, texture, and so on), at least if we share the artist's culture. Thus, a picture (or any other kind of artwork) is like a chair; a chair *can* be stood on or burned for firewood or used as a weapon, but it was created with a specific purpose that was evident and remains evident to all competent viewers—in this case people who are familiar with chairs. Further, it can be evaluated with reference to its purpose—we can say, for instance, that it is a poor chair because it is uncomfortable and fragile.

Formal Analysis Versus Description

Is the term *formal analysis* merely a pretentious substitute for *descrip-tion?* Not quite. A **description** is an impersonal inventory, dealing with the relatively obvious, reporting what any eye might see: "A woman in a white dress sits at a table, reading a letter. Behind her . . . ." It can also comment on the execution of the work ("thick strokes of paint," "a highly polished surface"), but it does not offer inferences, and it does not evalu-ate. A highly detailed description that seeks to bring the image before the reader's eyes—a kind of writing fairly common in the days before illustra-tions of artworks were readily available in books—is sometimes called an *ekphrasis* or *ecphrasis,* from the Greek word for "description" (*ek* = out, *phrazein* = tell, declare). Such a description may be set forth in terms that also seek to convey the writer's emotional response to the work. That is, the description praises the work by seeking to give the reader a sense of being in its presence, especially by commenting on the presumed emotions expressed by the depicted figures. Here is an example: "We recoil with the terrified infant, who averts his eyes from the soldier whose heart is as hard as his burnished armor."

Writing of this sort is no longer common; a description today is more likely to tell us, for instance, that the head of a certain portrait sculpture "faces front; the upper part of the nose and the rim of the right earlobe are missing. . . . The closely cropped beard and mustache are indicated by short random strokes of the chisel," and so forth. These statements, from an entry in the catalog of an exhibition, are all true and they can be useful, but they scarcely reveal the thought, the reflectiveness, that we associate with analysis. When the entry in the catalog goes on, however, to say that "the surfaces below the eyes and cheeks are sensitively modeled to suggest the soft, fleshly forms of age," we begin to feel that now indeed we are reading an analysis, because here the writer is arguing a thesis.

Similarly, although the statement that "the surface is in excellent condition" is purely descriptive (despite the apparent value judgment in "excellent"), the statement that the "dominating block form" of the portrait contributes to "the impression of frozen tension" can reasonably be called analytic. One reason we can characterize this statement as analytic is that it offers an argument, in this instance an argument concerned with cause and effect: The dominating block form produces an effect—*causes* us to perceive a condition of frozen tension.

Much of any formal analysis will inevitably consist of description ("The pupils of the eyes are turned upward"), and accurate descriptive writing itself requires careful observation of the object and careful use of words. But an essay is a formal analysis rather than a description only if it connects causes with effects, thereby showing *how* the described object works. For example, "The pupils of the eyes are turned upward" is a description, but the following revision is an analytic statement: "The pupils of the eyes are turned upward, suggesting a heaven-fixed gaze, or, more bluntly, suggesting that the figure is divinely inspired."

Another way of putting it is to say that analysis tries to answer the somewhat odd-sounding question, "*How* does the work mean?" Thus, the following paragraph, because it is concerned with *how* form makes meaning, is chiefly analytic rather than descriptive. The author has made the point that a Protestant church emphasizes neither the altar nor the pulpit; "as befits the universal priesthood of all believers," he says, a Protestant church is essentially an auditorium. He then goes on to analyze the ways in which a Gothic cathedral says or means something very different:

The focus of the space on the interior of a Gothic cathedral is . . . compulsive and unrelievedly concentrated. It falls, and falls exclusively, upon the sacrifice that is re-enacted by the mediating act of priest before the altar-table. So therefore, by design, the first light that strikes the eye, as one enters the cathedral, is the jeweled glow of the lancets in the apse, before which the altar-table stands. The pulsating rhythm of the arches in the nave arcade moves toward it; the string-course moldings converge in perspective recession upon it. Above, the groins of the apse radiate from it; the ribshafts which receive them and descend to the floor below return the eye inevitably to it. It is the single part of a Gothic space in which definiteness is certified. In any other place, for any part which the eye may reach, there is always an indefinite beyond, which remains to be explored. Here there is none. The altar-table is the common center in which all movement comes voluntarily to rest.

—John F. A. Taylor, *Design and Expression in the Visual Arts*
(New York: Dover, 1964), 115–17

In this passage the writer is telling us, analytically, *how* the cathedral means.

Opposition to Formal Analysis

Formal analysis, we have seen, assumes that artists shape their materials so that a work of art embodies a particular meaning and evokes a pleasurable response in the spectator. The purpose of formal analysis is to show *how* intended meanings are communicated. Since about 1970, however, these assumptions have been strongly called into question. There has been a marked shift of interest from the work as a thing whose meaning is contained within itself—a decontextualized object—to a thing whose meaning partly, largely, or even entirely consists of its context, its relation to things outside of itself (for instance, the institutions or individuals for whom the work was produced), especially its relationship to the person who perceives it. Works of art, many people now say, do not speak for themselves. Viewers speak for the works, performing acts of ventriloquism. That is, viewers are not innocent or neutral eyes, passively receiving the meaning that the artist put into the work, but rather they are persons who put meaning—depending on their experience—into the work they look at. (This point has been discussed on pages 16–21.)

Further, there has been a shift from viewing an artwork as a thing of value in itself—or as an object that provides pleasure and that conveys some sort of profound and perhaps universal meaning—to viewing the artwork as an object that reveals the power structure of a society. The work is brought down to earth, so to speak, and is said thereby to be "demystified." Thus the student does not look for a presumed unified whole. On the contrary, the student "deconstructs" the work by looking for "fissures" and "slippages" that give away—reveal, unmask—the underlying political and social realities that the artist sought to cover up with sensuous appeal.

A discussion of an early nineteenth-century idyllic landscape painting, for instance, today might call attention not to the elegant brushwork and the color harmonies (which earlier might have been regarded as sources of aesthetic pleasure), or even to the neat hedges and meandering streams (meant to evoke pleasing sensations), but to such social or psychological matters as the painter's unwillingness to depict the hardships of rural life and the cruel economic realities of land ownership in an age when poor families could be driven from their homes at the whim of a rich landowner. Such a discussion might even argue that the picture, by means of its visual seductiveness, seeks to legitimize social inequities. (We will return to the matters of demystification and deconstruction in Chapter 6, when we look at the social historian's approach to artworks, on pages 151–56.)

We can grant that works of art are partly shaped by social and political forces (these are the subjects of historical and political approaches, discussed in Chapter 6); and we can grant that works of art are partly shaped by the artist's personality (the subject of psychoanalytical approaches, also discussed in Chapter 6). But this is only to say that works of art can be studied from several points of view; it does not invalidate the view that these works are also, at least in part, shaped by conscious intentions, and that the shapes or constructions that the artists (consciously or not) have produced convey a meaning.

STYLE AS THE SHAPER OF FORM

It is now time to define the elusive word **style.** The first thing to say is that the word is *not* used by most art historians to convey praise, as in "He has style." Rather, it is used neutrally, for everyone and everything made has a style—good, bad, or indifferent. The person who, as we say, "talks like a book" has a style (probably an annoying one), and the person who keeps saying "Uh, you know what I mean" has a style too (different, but equally annoying).

Similarly, whether we wear jeans or painter's pants or gray flannel slacks, we have a style in our dress. We may claim to wear any old thing, but in fact we don't; there are clothes we wouldn't be caught dead in. The clothes we wear are expressive; they announce that we are police officers or bankers or tourists or college students—or at least they show what we want to be thought to be, as when in the 1960s many young middle-class students wore tattered clothing, thus showing their allegiance to the poor and their enmity toward what was called the Establishment. It is not silly to think of our clothing as a sort of art that we make. Once we go beyond clothing as something that merely serves the needs of modesty and that provides protection against heat and cold and rain, we get clothing whose style is expressive.

To turn now to our central topic—style in art—we can all instantly tell the difference between a picture by van Gogh and one by Norman Rockwell or Walt Disney, even though the subject matter of all three pictures may be the same (e.g., a seated woman). How can we tell? By the style—that is, by line, color, medium, and all of the other things we talked about earlier in this chapter. Walt Disney's figures tend to be built up out of circles and ovals (think of Mickey Mouse), and the color shows no modeling or traces of brush strokes; Norman Rockwell's methods of depicting figures are different, and van Gogh's are different in yet other ways. Similarly, a Chinese landscape, painted with ink on silk or on paper, simply cannot look like a van Gogh landscape done with oil paint on canvas, partly because the materials prohibit such identity and partly because the Chinese painter's vision of landscape (usually lofty mountains) is not van Gogh's vision. Their works "say" different things. As the poet Wallace Stevens put it, "A change of style is a change of subject."

We recognize certain *distinguishing characteristics* (from large matters, such as choice of subject and composition, to small matters, such as kinds of brush strokes) that mark an artist, or a period, or a culture, and these constitute the style. Almost anyone can distinguish between a landscape painted by a traditional Chinese artist and one painted by van Gogh. But it takes considerable familiarity with van Gogh to be able to say of a work, "Probably 1888 or maybe 1889," just as it takes considerable familiarity with the styles of Chinese painters to be able to say, "This is a Chinese painting of the seventeenth century, in fact the late seventeenth century. It belongs to the Nanking School and is a work by Kung Hsien—not by a follower, and certainly not a copy, but the genuine article."

Style, then, is revealed in **form;** an artist creates form by applying certain techniques to certain materials, in order to embody a particular vision or content. In different ages people have seen things differently: the nude body as splendid, or the nude body as shameful; Jesus as majestic ruler, or Jesus as the sufferer on the cross; landscape as pleasant, domesticated countryside, or landscape as wild nature. So the chosen subject matter is not only part of the content but is also part of that assemblage of distinguishing characteristics that constitutes a style.

All of the elements of style, finally, are expressive. Take ceramics as an example. The kind of clay, the degree of heat at which it is baked, the decoration or glaze (if any), the shape of the vessel, the thickness of its wall, all are elements of the potter's style, and all contribute to the expressive form. But every expressive form is not available in every age; certain visions, and certain technologies, are, in certain ages, unavailable. Porcelain, as opposed to pottery, requires a particular kind of clay and an extremely high temperature in the kiln, and these were simply not available to the earliest Japanese potters. Even the potter's wheel was not available to them; they built their pots by coiling ropes of clay and then, sometimes, they smoothed the surface with a spatula. The result is a kind of thick-walled, low-fired ceramic that expresses energy and earthiness, far different from those delicate Chinese porcelains that express courtliness and the power of technology (or, we might say, of art).

SAMPLE ESSAY: A FORMAL ANALYSIS

The following sample essay, written by an undergraduate, includes a good deal of description (a formal analysis usually begins with a fairly full description of the artwork), and the essay is conspicuously impersonal (another characteristic of a formal analysis). But notice that even this apparently dispassionate assertion of facts is shaped by a **thesis.** If we stand back from the essay, we can see that the basic argument is this: The sculpture successfully combines a highly conventional symmetrical style, on the one hand, with mild asymmetry and a degree of realism, on the other.

Put thus, the thesis does not sound especially interesting, but that is because the statement is highly abstract, lacking in concrete detail. A writer's job is to take that idea (thesis) and to present it in an interesting and convincing way. The idea will come alive for the reader when the writer supports it by calling attention to specific details—the evidence—as the student writer does in the following essay.

Egyptian, *Seated Statue of Prince Khunera as a Scribe,* 2548–2524 BC. Yellow limestone, 12″. (Courtesy of Museum of Fine Arts, Boston, Harvard-Boston Expedition)

Stephen Beer

Fine Arts 10A

September 10, 1999

Formal Analysis: Prince Khunera as a Scribe

Prince Khunera as a Scribe, a free-standing Egyptian sculpture 12 inches tall, now in the Museum of Fine Arts, Boston, was found at Giza in a temple dedicated to the father of the prince, King Mycerinus. The limestone statue may have been a tribute to that

Fourth Dynasty king.[1] The prince, sitting cross-legged with a scribal tablet on his lap, rests his hands on his thighs. He wears only a short skirt or kilt.

The statue is in excellent condition although it is missing its right forearm and hand. Fragments of the left leg and the scribe's tablet also have been lost. The lack of any difference in the carving between the bare stomach and the kilt suggests that these two features were once differentiated by contrasting paint that has now faded, but the only traces of paint remaining on the figure are bits of black on the hair and eyes.

The statue is symmetrical, composed along a vertical axis which runs from the crown of the head into the base of the sculpture. The sculptor has relied on basic geometric forms in shaping the statue on either side of this axis. Thus, the piece could be described as a circle (the head) within a triangle (the wig) which sits atop a square and two rectangles (the torso, the crossed legs, and the base). The reliance on basic geometric forms reveals itself even in details. For example, the forehead is depicted as a small triangle within the larger triangular form of the headdress.

On closer inspection, however, one observes that the rigidity of both this geometric and symmetric organization is relieved by the artist's sensitivity to detail and by his ability as a sculptor. None of the shapes of the work is a true geometric form. The full, rounded face is more of an oval than a circle, but actually it is neither. The silhouette of the upper part of the body is defined by softly undulating lines that represent the muscles of the arms and that modify the simplicity of a strictly square shape. Where the

---

[1]Museum label.

prince's naked torso meets his kilt, just below the waist, the
sculptor has suggested portliness by allowing the form of the
stomach to swell slightly. Even the "circular" navel is flattened into
an irregular shape by the suggested weight of the body. The
contours of the base, a simple matter at first glance, actually are not
exactly four-square but rather are slightly curvilinear. Nor is the
symmetry on either side of the vertical axis perfect: Both the mouth
and the nose are slightly askew; the right and left forearms
originally struck different poses; and the left leg is given
prominence over the right. These departures from symmetry and
from geometry enliven the statue, giving it both an individuality
and a personality.

Although most of the statue is carved in broad planes, the
sculptor has paid particular attention to details in the head. There
he attempted to represent realistically the details of the prince's
face. The parts of the eye, for example--the eyebrow, eyelids,
eyeballs, and sockets--are distinct. Elsewhere the artist has not
worked in such probing detail. The breasts, for instance, are
rendered in large forms, the nipples being absent. The attention to
the details of the face suggests that the artist attempted to render a
lifelikeness of the prince himself.

The prince is represented in a scribe's pose but without a
scribe's tools. The prince is not actually doing anything. He merely
sits. The absence of any open spaces (between the elbows and the
waist) contributes to the figure's composure or self-containment.
But if he sits, he sits attentively: There is nothing static here. The
slight upward tilt of the head and the suggestion of an upward gaze
of the eyes give the impression that the alert prince is attending
someone else, perhaps his father the king. The suggestion in the
statue is one of imminent work rather than of work in process.

Thus, the statue, with its variations from geometric order, suggests the presence, in stone, of a particular man. The pose may be standard and the outer form may appear rigid at first, yet the sculptor has managed to depict an individual. The details of the face and the overfleshed belly reveal an intent to portray a person, not just an idealized member of the scribal profession. Surely when freshly painted these elements of individuality within the confines of conventional forms and geometric structure were even more compelling.

Behind the Scene: Beer's Essay, from Early Responses to Final Version

Let's go backstage, so to speak, to see how Stephen Beer turned his notes into a final draft.

Beer's Earliest Responses. After studying the object and reading the museum label, Beer jotted down some ideas on 4 × 6-inch index cards. What historical information does the label provide? (Beer recorded the material on a note card.) Can the sculpture be called realistic? Yes and no. (Beer put his responses, in words, on the cards.) What is the condition of the piece? (Again he put his responses into writing, on cards.) A day later, when he returned to work on his paper, stimulated by another look at the artwork and by a review of his notes, Beer made additional jottings.

Organizing Notes. When the time came to turn the notes into a draft and the revised draft into an essay, Beer reviewed the cards and he added further thoughts. Next, he organized the note cards, putting together into a packet whatever cards he had about (for instance) realism, and putting together, into another packet, whatever cards he had about (again, for instance) background material. Reviewing the cards in each packet, and on the basis of this review discarding a few cards that no longer seemed useful and moving an occasional card to a different packet, Beer started to think about how he might organize his essay.

As a first step in settling on an organization, he arranged the packets into a sequence that seemed reasonable to him. It made sense, he thought, to begin with some historical background and a brief description, then to touch on Egyptian sculpture in general (but he soon decided *not* to include this general material), then to go on to some large points about the particular piece, then to refine some of these points, and finally to offer some sort of conclusion. This organization, he felt, was reasonable and would enable his reader to follow his argument easily.

Preparing a Preliminary Outline. In order to get a clearer idea of where he might be going, Beer then keyboarded—following the sequence of his packets—the gist of what at this stage seemed to be the organization of his chief points. In short, he prepared a map or rough outline so that he could easily see, almost at a glance, if each part of his paper would lead coherently to the next part.

Beer realized that in surveying his outline he might become aware of points that he should have included but that he had instead overlooked.

A tentative outline, after all, is not a straitjacket that determines the shape of the final essay. To the contrary, it is a preliminary map that, when examined, helps the writer to see not only pointless detours—these will be eliminated in the draft—but also undeveloped areas that need to be worked up. As the handwritten additions in Beer's outline indicate, after drafting his map he made some important changes before writing a first draft:

Background: Egypt to Boston;
   free-standing, 12″ limestone;
    Giza overall view: sitting, hands on thighs
~~General remarks on Egyptian sculpture~~
condition:
   some parts missing
   paint almost all gone
symmetry (geometry?)   *description*
   square (body)
    circle in a triangle (head)
    two rectangles
variations in symmetry
   body varied   *face not really a circle (oval)*
   smaller variations (face, ^navel, forearms)
realism (yes and no)   *Pose:*
conclusion, summary (?):   *Static? Active? Both?*
   geometric but varied
   individualized (??)
   combination

Writing a Draft. Working from his thoughtfully revised outline, Beer wrote a first draft, which he then revised into a second draft. The second draft, when further revised, became the final essay.

The word *draft,* by the way, comes from the same Old English root that gives us *draw.* When you draft an essay, you are making a sketch, putting on paper a sketch or plan that you will later refine.

Outlining a Draft. A good way to test the coherence of a final draft—to see if indeed it qualifies as an essay rather than as a draft—is to outline it, paragraph by paragraph, in *two* ways, indicating

(a) what each paragraph *says*
(b) what each paragraph *does*

Here is a double outline of this sort for Beer's seven-paragraph essay. In (a) we get a brief summary of what Beer is *saying* in each paragraph, and in (b) we get, in the italicized words, a description of what he is *doing* in the paragraph.

1. a. Historical background and brief description
    b. *Introduces* the artwork
2. a. The condition of the artwork
    b. *Provides further overall description, preparatory* to the analysis
3. a. The geometry of the work
    b. *Introduces the thesis,* concerning the basic, overall geometry
4. a. Significant details
    b. *Modifies (refines) the argument*
5. a. The head
    b. *Compares* the realism of head with the breasts, in order to make the point that the head is more detailed
6. a. The pose
    b. *Argues* that the pose is not static
7. a. Geometric, yet individual
    b. *Concludes,* largely by *summarizing*

An outline of this sort, in which you force yourself to consider not only the content but also the function of each paragraph, will help you to see if your essay (a) says something and (b) says it with the help of an effective structure. If the structure is sound, your argument will flow nicely.

POSTSCRIPT: THOUGHTS ABOUT THE
WORDS "REALISTIC" AND "IDEALIZED"

In his fifth paragraph (page 90) Beers uses the word "realistically," and in his final paragraph he uses "idealized." These words, common in essays on art, deserve comment. Let's begin a bit indirectly. Aristotle (384–322 BC) says that the arts originate in two basic human impulses, the impulse to imitate (from the Greek word *mimesis,* imitation) and the impulse to create patterns or harmony. In small children we find both, (1) the impulse to imitate in their mimicry of others and (2) the impulse to harmony in their fondness for rocking and for strongly rhythmic nursery rhymes. Most works of art, as we shall see, combine imitation (mimicry, a representation of what the eye sees, realism) with harmony (an overriding form or pattern produced by a shaping idea). "We can imagine," Kenneth Clark wrote, "that the early sculptor who found the features of a head conforming to an ovoid, or the body conforming to a column, had a deep satisfaction. Now it looks as if it would last" (Introduction to *Masterpieces of Fifty Centuries,* 1970, page 15).

For an extreme example of a body simplified to a column—a body shaped, in Clark's words, by the *idea* that the body conforms to a column—we can look at Constantin Brancusi's *Torso of a Young Man* (1924) (opposite). Here the artist's *idea* has clearly dominated his eye; this body is **idealized.** Looking at this work, we are not surprised to learn that Brancusi said he was concerned with the "eternal type [i.e., the prototype] of ephemeral forms," and that "What is real is not the external form, but the essence of things. . . . It is impossible for anyone to express anything essentially real by imitating its exterior surface." The real or essential form represented in this instance is both the young man of the title and also the phallus.

Less extreme examples of idealized images are provided by many portrait heads of Antinous (also Antinoös), the youth beloved by the Roman emperor Hadrian. Writing in the *Bulletin* of the Metropolitan Museum of Art, Elizabeth J. Milleker calls attention to the combination of "actual features of the boy" and "an idealized image" in such a head (see p. 96):

> This head is a good example of the sophisticated portrait type created by imperial sculptors to incorporate what must have been actual features of the boy in an idealized image that conveys a godlike beauty. The ovoid face with a straight brow, almond-shaped eyes, smooth cheeks, and fleshy lips is surrounded by abundant tousled curls. The ivy

Constantin Brancusi, *Torso of a Young Man*, 1924. Polished brass, 18″; with original wood base 58½″. (The Hirshhorn Museum and Sculpture Garden, Smithsonian Institution, gift of Joseph H. Hirshhorn, 1966. Photographer, Lee Stalsworth)

wreath encircling his head associates him with Dionysos, a guarantor of renewal and good fortune.
—Metropolitan Museum of Art *Bulletin* (Fall 1997): 15

Realism has at least two meanings in writings about art: (1) a movement in mid-nineteenth-century Western Europe and America, which emphasized the everyday subjects of ordinary life, as opposed to subjects

Roman, *Portrait Head of Antinoös,* AD 130–38. Marble, 9¾". (The Metropolitan Museum of Art. Gift of Bronson Pinchot, in recognition of his mother, Rosina Asta Pinchot, 1996. 1996.401)

drawn from mythology, history, and upperclass experience; (2) fidelity to appearances, the accurate rendition of the surfaces of people, places, and things. In our discussion of realism, we will be concerned only with the second definition. *Naturalism* is often used as a synonym for *realism;* thus, a work that reproduces surfaces may be said to be realistic, naturalistic, or illusionistic; *veristic* is also used, but less commonly. The most extreme form of realism is *trompe-l'oeil* (French: deceives the eye), complete illusionism—the painted fly on the picture frame, the waxwork museum guard standing in a doorway, images created with the purpose of deceiving the viewer. But of course most images are not the exact size of the model, so even if they are realistically rendered they do not deceive. When we look at most images, we are aware that we are looking not at reality (a fly, a human being) but at the product of the artist's gaze at such real things. Further, the medium itself may prohibit illusionism, an unpainted stone or bronze head, however accurate in its representation of cheekbones, hair, the shape of the nose, and so forth, cannot be taken for Abraham Lincoln.

Idealism, like realism, has at least two meanings in art: (1) the belief that a work conveys an idea as well as appearances and (2) the belief—

derived partly from the first meaning—that it should convey an idea that elevates the thoughts of the spectator, and it does this by presenting an image, let's say of heroism or of motherhood, loftier than any real object that we can see in the imperfect world around us. (Do not confuse *idealism* as it is used in art with its everyday meaning, as in "despite her years, she retained her idealism," where the word means "noble goals.") We know that in the sixteenth century Raphael, searching for a model for his painting of Galatea, could not find a woman who fulfilled his "idea" of beauty, and so, he wrote to a friend, he had to draw upon parts of various women to paint an image that expressed the ideal woman. Somewhat similarly, the Hadrianic sculptors who made images of Antinous, as the writer in the Metropolitan Museum's *Bulletin* said, must have had in mind not a particular youth but the idea of "godlike beauty."

By way of contrast, consider Sir Peter Lely's encounter with Oliver Cromwell (1599–1658), the English general and statesman. Cromwell is reported to have said to the painter, "Mr. Lely, I desire you would use all your skill to paint my picture truly like me, and not flatter me at all; but remark [i.e., take notice of] all these roughnesses, pimples, warts, and everything as you see me." Cromwell was asking for a realistic (or naturalistic, or veristic) portrait, not an idealized portrait like the sculpture of Antinous. What would an idealized portrait look like? It would omit the blemishes. Why? Because the blemishes would be thought to be mere trivial surface details that would get between the viewer and the artist's *idea* of Cromwell, Cromwell's essence as the artist perceives it—for instance, the nobility that characterized his statesmanship and leadership. What distinguishes the idealizing artist from the ordinary person, it is said, is the artist's imaginative ability to penetrate the visible (the surface) and set forth an elevating ideal.

In short, *realism* is defined as the representation of visual phenomena as exactly—as realistically—as the medium (stone, bronze, paint on paper or canvas) allows. At the other extreme from illusionism we have idealism, for instance in the representation of the torso of a young man by a column. A realistic portrait of Cromwell will show him as he appears to the eye, warts and all; an idealized portrait will give us the idea of Cromwell by, so to speak, airbrushing the warts, giving him some extra stature, slimming him down a bit, giving him perhaps a more thoughtful face than he had, setting him in a pyramidal composition to emphasize his stability, thereby stimulating our minds to perceive the nobility of his cause.

Both realism and idealism have had their advocates. As a spokesperson for realism we can take Leonardo, who in his *Notebooks* says that "the mind of the painter should be like a looking-glass that is filled with as many images as there are objects before him." Against this view we can take a remark by a contemporary painter, Larry Rivers: "I am not interested in the art of holding up mirrors." Probably most artists offer the Aristotelian combination of imitation and harmony. The apparently realistic (primarily mimetic) artist is concerned at least in some degree with a pattern or form that helps to order the work and to give it meaning, and the apparently idealistic artist—even the nonobjective artist who might seem to deal only in harmonious shapes and colors—is concerned with connecting the work to the world we live in, for instance, to our emotions. An artist might deliberately depart from surface realism—mimetic accuracy—in order to "defamiliarize" or "estrange" our customary perceptions, slowing us down or shaking us up, so to speak, in order to jostle us out of our stock responses, thereby letting us see reality freshly. Although this idea is especially associated with the Russian Formalist school of the early twentieth century, it can be traced back to the early nineteenth-century Romantic writers. For instance, Samuel Taylor Coleridge praised the poetry of William Wordsworth because, in Coleridge's words, it removed "the film of familiarity" that clouded our usual vision.

Somewhere near the middle of the spectrum, between artists who offer highly mimetic representations and those who offer representations that bear little resemblance to what we see, we have the sculptor, hypothesized by Kenneth Clark, who saw the head as an ovoid and said to himself, "Now it looks as if it would last." Here the "idea" that shapes the features of the head (for instance, bringing the ears closer to the skull) is the idea of perfection and endurance, stability, even eternity, and surely some such thoughts cross our minds when we perceive works that we love. One might almost write a history of art in terms of the changing proportions of realism and idealism during the lifetime of a culture.

It is easy to find remarks by artists setting forth a middle view. In an exhibition catalog (1948), Henri Matisse said, "There is an inherent truth which must be disengaged from the outward appearances of the object to be represented. . . . Exactitude is not truth." And most works of art are neither purely realistic (concerned only with "exactitude," realistic description) nor purely idealistic (concerned only with "an inherent truth"). Again we think of Aristotle's combination of the impulse to imitate and the impulse to create harmony. We might think, too, of a comment by Picasso: "If you want to draw, you must first shut your eyes and sing." We can probably agree that *Prince Khunera as a Scribe* (page 88) shows a good deal of idealism, but it also shows realistic touches. Stephen Beer's analysis calls attention to its idealized quality, in its symmetry and its nearly circular head, but he also says that the eye is rendered with "de-

> ✍ **A RULE FOR WRITERS:**
>
> There is nothing wrong with using the words *realistic* and *idealized* in your essay, but keep in mind that it is not a matter of all-or-nothing; there are degrees of realism and degrees of idealization

scriptive accuracy." Or look at Michelangelo's *David* (page 28). It is realistic in its depiction of the veins in David's hands, but it is idealized in its color (not flesh color but white to suggest purity), in its size (much larger than life, to convey the ideal of heroism), and in its nudity. Surely Michelangelo did not think David went into battle naked, so why is his David nude? Because Michelangelo, carving the statue in part to commemorate the civic constitution of the Florentine republic, wanted to convey the ideas of justice and of classical heroism, and classical sculptures of heroes were nude. We can, then, talk about Michelangelo's idealism, and—still talking of the same image—we can talk about his realism.

Consider the three pictures of horses shown on page 100. Han Gan's painting (upper left) is less concerned with accurately rendering the appearance of a horse than with rendering its great inner spirit, hence the head and neck that are too large for the body, and the body that is too large for the legs. The legs, though in motion to show the horse's grace and liveliness, are diminished because the essence of the horse is the strength of its body, communicated partly by juxtaposing the arcs of its rump and its (unreal) electrified mane with the stolid hitching post. In brief, the painting shows what we in the West probably would call an idealized horse, although the Chinese might say that by revealing the spirit the picture captures the "real" horse. (It once had a tail, but the tail has been largely eroded by wear, and the vestiges have been obscured by an owner's seal.)

Leonardo's drawing (upper right) is largely concerned with anatomical correctness, and we can call it realistic. Still, by posing the horse in profile Leonardo calls attention to the animal's geometry, notably the curves of the neck, chest, and rump, and despite the accurate detail the picture seems to represent not a particular horse but the essential idea of a horse. (Doubtless the blank background and the absence of a groundplane here, as in the Chinese painting, contribute to this impression of idealizing.)

George Stubbs's painting of Hambletonian (bottom), who had recently won an important race, surely is an accurate representation of a particular horse, but even here we can note an idealizing element: Stubbs emphasizes the animal's heroic stature by spreading its image across the canvas so that the horse dwarfs the human beings and the buildings.

Upper left: Han Gan, *Night-Shining White,* mid-eighth century. Ink on paper, 11¹³⁄₁₆″ high. (The Metropolitan Museum of Art, New York. The Dillon Fund, 1977. 1977.78)
Upper right: Leonardo da Vinci, *A Horse in Profile to the Right, and Its Forelegs,* c. 1490. Silverpoint on blue prepared surface, 8½″ × 6⅜″. (The Royal Collection. © Her Majesty Queen Elizabeth II)
Bottom: George Stubbs, *Hambletonian,* 1800. Oil on canvas, 82½″ × 144⅝″. (Mount Stewart House and Garden/National Trust Photographic Library)

If your instructor asks you to compare two works—perhaps an Egyptian ruler and a Greek athlete, or an Indian Buddha and a Chinese Buddha—you may well find one of them more realistic than the other, but remember, even a highly realistic work may include idealized elements, and an idealized work may include realistic elements.

3

Writing a Comparison

COMPARING AS A WAY OF DISCOVERING

Analysis frequently involves comparing: Things are examined for their resemblances to and differences from other things. Strictly speaking, if one emphasizes the differences rather than the similarities, one is contrasting rather than comparing, but we need not preserve this distinction; we can call both processes *comparing.*

Although your instructor may ask you to write a comparison of two works of art, the *subject* of the essay is the *works,* or, more precisely, the subject is the thesis you are advancing; for example, that one work is later than the other or is more successful. Comparison is simply an effective analytical *technique* to show some of the qualities of the works. We usually can get a clearer idea of what X is when we compare it to Y—provided that Y is at least somewhat like X. Comparing, in short, is a way of discovering, a way of learning.

In the words of Howard Nemerov, "If you really want to see something, look at something else." But the "something else" can't be any old thing. It has to be relevant. For example, in a course in architecture you may compare two subway stations (considering the efficiency of the pedestrian patterns, the amenities, and the aesthetic qualities), with the result that you may come to understand both of them more fully; but a comparison of a subway station with a dormitory, no matter how elegantly written, can hardly teach the reader or the writer anything. Nor can a comparison of the House of Commons with the House of Pancakes teach anything, as Judith Stone entertainingly demonstrates in an article in *The New York Times Magazine* (26 February 1995): She compares the number (1 House of Commons versus 620 Houses of Pancakes), the tipping practice (not allowed in the House of Commons but permitted in the House of Pancakes), the preferred salutation ("My honorable friend" versus "hon"), and so forth. If you keep in mind the principle that a comparison should help you to learn, you will not make useless comparisons.

Art historians almost always use comparisons when they discuss authenticity: A work of uncertain attribution is compared with undoubtedly genuine works on the assumption that the uncertain work, when closely compared with genuine works, will somehow be markedly different, perhaps in brush technique, and thereby shown probably not to be genuine (here we get to the thesis) despite superficial similarities of, say, subject matter and medium. (This assumption can be challenged—a given artist may have produced a work with unique characteristics—but it is nevertheless widely held.)

Comparisons are also commonly used in dating a work, and thus in tracing the history of an artistic movement or the development of an artist's career. The assumption here is that certain qualities in a work indicate the period, the school, perhaps the artist, and even the period within the artist's career. Let's assume, for instance, that there is no doubt about who painted a particular picture, and that the problem is the date of the work. By comparing this work with a picture that the artist is known to have done, say, in 1850, and with yet another that the artist is known to have done in 1870, one may be able to conjecture that the undated picture was done, say, midway between the dated works, or that it is close in time to one or the other.

The assumptions underlying the uses of comparison are that an expert can recognize not only the stylistic characteristics of an artist, but can also identify those that are permanent and can establish the chronology of those that are temporary. In practice these assumptions are usually based on yet another assumption: A given artist's early works are relatively immature; the artist then matures, and if there are some dated works, we can with some precision trace this development or evolution. Whatever the merits of these assumptions, comparison is a tool by which students of art often seek to establish authenticity and chronology.

✍ A RULE FOR WRITERS:

The point of a comparison is not simply to list similarities or differences; the point is to make a point.

TWO WAYS OF ORGANIZING A COMPARISON

We can call the two ways of organizing a comparison *block-by-block* (or, less elegantly but perhaps more memorably, *lumping*) and *point-by-point* (or *splitting*). When you compare block-by-block you say what you have to say about one artwork in a block or lump, and then you go on to discuss the second artwork, in another block or lump. When you compare point-by-point, however, you split up your discussion of each work, more or less interweaving your comments on the two things being compared, perhaps in alternating paragraphs, or even in alternating sentences.

Here is a miniature essay—it consists of only one paragraph—that illustrates lumping. The writer compares a Japanese statue of a Buddha (below) with a Chinese statue of a bodhisattva (p. 104). (A Buddha has achieved enlightenment and has withdrawn from the world. A bodhisattva—in Sanskrit the term means "enlightened being"—is, like a Buddha, a person of very great spiritual enlightenment, but unlike a Bud-

Shaka nyorai,
Sakyamuni Buddha,
Japanese, late tenth
century. Wood, 33½".
(Denman Waldo
Ross Collection;
courtesy of Museum
of Fine Arts, Boston)

Bodhisattva Kuan Yin Seated in Royal Ease Position, Chinese, twelfth century. Carved wood, 56½″. (Harvey Edward Wetzel Fund; courtesy of Museum of Fine Arts, Boston)

dha, a bodhisattva chooses to remain in this world in order to save humankind.) The writer's point here is simply to inform the museum-goer that all early East Asian religious images are not images of the Buddha. The writer says what she has to say about the Buddha, all in one lump, and then in another lump says what she has to say about the bodhisattva.

The Buddha, recognizable by a cranial bump that indicates a sort of supermind, sits erect and austere in the lotus position (legs crossed, each foot with the sole upward on the opposing thigh), in full control of his body. The carved folds of his garments, in keeping with the erect posture, are severe, forming a highly disciplined pattern that is an outward expression of his remote, constrained, austere inner nature. The bodhisattva, on the other hand, sits in a languid, sensuous posture known as "royal ease," the head pensively tilted downward, one knee elevated, one leg hanging down. He is accessible, relaxed, and compassionate.

The structure is, simply, this:

The Buddha (posture, folds of garments, inner nature)

The bodhisattva (posture, folds of garments, inner nature)

If, however, the writer had wished to split rather than to lump, she would have compared an aspect of the Buddha with an aspect of the bodhisattva, then another aspect of the Buddha with another aspect of the bodhisattva, and so on, perhaps ending with a synthesis to clarify the point of the comparison. The paragraph might have read like this:

> The Buddha, recognizable by a cranial bump that indicates a sort of supermind, sits erect and austere, in the lotus position (legs crossed, each foot with the sole upward on the opposing thigh), in full control of his body. In contrast, the bodhisattva sits in a languid, sensuous posture known as "royal ease," the head pensively tilted downward, one knee elevated, one leg hanging down. The carved folds of the Buddha's garments, in keeping with his erect posture, are severe, forming a highly disciplined pattern, whereas the bodhisattva's garments hang naturalistically. Both figures are spiritual but the Buddha is remote, constrained, and austere; the bodhisattva is accessible, relaxed, and compassionate.

In effect the structure is this:

The Buddha (posture)

The bodhisattva (posture)

The Buddha (garments)

The bodhisattva (garments)

The Buddha and the bodhisattva (synthesis)

When you offer an extended comparison, it is advisable to begin by defining the main issue or problem—for instance, the kind of ivory, the subject matter, the treatment of space, and the style of the carving suggest that this piece is fourteenth-century French and that piece is a modern fake—and also by indicating what your principle of organization will be. *Caution:* an essay that uses splitting too rigidly is likely to produce a

Ping-Pong effect. There is also the danger that the essay will not come into focus—the point will not be grasped—until the essayist stands back from the seven-layer cake and announces, in the concluding paragraph, that the odd layers taste better. In one's preparatory thinking one may employ splitting in order to get certain characteristics clear in one's mind, but one must come to some conclusions about what these add up to *before* writing the final version.

The final version should not duplicate the thought processes; rather, since the point of a comparison is *to make a point,* it should be organized so as to make the point clearly and effectively. Lumping will often do the trick. After reflection you may decide that although there are superficial similarities between X and Y, there are essential differences; in the finished essay, then, you probably will not wish to obscure the main point by jumping back and forth from one work to the other, working through a series of similarities and differences. It may be better to announce your thesis, then discuss X, and then Y.

Whether in any given piece of writing you should compare by lumping or by splitting will depend largely on your purpose and on the complexity of the material. Lumping is usually preferable for long, complex comparisons, if for no other reason than to avoid the Ping-Pong effect, but no hard-and-fast rule covers all cases here. Some advice, however, may be useful:

If you split, in rereading your draft:

- *Ask yourself if your imagined reader can keep up with the back-and-forth movement.* Make sure (perhaps by a summary sentence at the end) that the larger picture is not obscured by the zigzagging.
- *Don't leave any loose ends.* Make sure that if you call attention to points 1, 2, and 3 in X, you mention all of them (not just 1 and 2) in Y.

If you lump, do not simply comment first on X and then on Y.

- *Let your reader know where you are going,* probably by means of an introductory sentence.
- *Don't be afraid in the second half to remind the reader of the first half.* It is legitimate, even desirable, to relate the second half of the comparison (chiefly concerned with Y) to the first half (chiefly concerned with X). Thus, you will probably say things like

"Unlike X, Y shows . . ." or "Although Y superficially resembles X in such-and-such, when looked at closely Y shows . . . ." In short, a comparison organized by lumping will not break into two separate halves if the second half develops by reminding the reader how it differs from the first half.

Again, the point of a comparison is to call attention to the unique features of something by holding it up against something similar but significantly different. If the differences are great and apparent, a comparison is a waste of effort. (Blueberries are different from elephants. Blueberries do not have trunks. And elephants do not grow on bushes.) Indeed, a comparison between essentially and obviously unlike things will merely confuse, for by making the comparison, the writer implies that there are significant similarities, and readers can only wonder why they do not see them. The essays that do break into unrelated halves are essays that make uninstructive comparisons: The first half tells the reader about five qualities in El Greco; the second half tells the reader about five different qualities in Rembrandt. You will notice in the student essay below that the second half occasionally looks back to the first half.

SAMPLE ESSAY: A STUDENT'S COMPARISON

This essay, by an undergraduate, discusses one object and then discusses a second. It lumps rather than splits. It does not break into two separate parts because at the start it looks forward to the second object, and in the second half of the essay it occasionally reminds us of the first object.

When you read this essay, don't let its excellence lead you into thinking that you can't do as well. The essay, keep in mind, is the product of much writing and rewriting. As Rebecca Bedell wrote, her ideas got better and better, for in her drafts she sometimes put down a point and then realized that it needed strengthening (e.g., with concrete details) or that—come to think of it—the point was wrong and ought to be deleted. She also derived some minor assistance—for facts, not for her fundamental thinking—from books, which she cites in footnotes.

Brief marginal annotations have been added to the following essay in order to help you appreciate the writer's skill in presenting her ideas.

Rebecca Bedell

FA 232 American Art

Title is focused and, in "Development," implies the thesis

John Singleton Copley's Early Development: From Mrs. Joseph Mann to Mrs. Ezekial Goldthwait

Opening paragraph is unusually personal but engaging, and it implies the problem the writer will address

Several Sundays ago while I was wandering through the American painting section of the Museum of Fine Arts, a professorial bellow shook me. Around the corner strode a well-dressed mustachioed member of the art historical elite, a gaggle of note-taking students following in his wake. "And here," he said, "we have John Singleton Copley." He marshaled his group about the rotunda, explaining that, "as one can easily

Thesis is clearly announced

see from these paintings, Copley never really learned to paint until he went to England."

A walk around the rotunda together with a quick leafing through a catalog of Copley's work should convince any viewer that Copley reached his artistic maturity years before he left for England in 1774. A comparison of two paintings at the Museum of Fine Arts, Mrs. Joseph Mann of 1753 and Mrs. Ezekial Goldthwait of ca. 1771, reveals that Copley had made huge advances in his style and technique even before he left America; by the time of his departure he was already a great portraitist. Both paintings are half-length portraits of seated women, and both are accompanied by paired portraits of their husbands.

Brief description of the first work

The portrait of Mrs. Joseph Mann, the twenty-two-year-old wife of a tavern keeper in Wrentham,

John Singleton Copley, *Mrs. Joseph Mann,* 1753. Oil on canvas, 36″ × 28¼″. (Gift of Frederick and Holbrook Metcalf; courtesy of Museum of Fine Arts, Boston)

John Singleton Copley, *Mrs. Ezekial Goldthwait,* ca. 1771. Oil on canvas, 50″ × 40″. (Bequest of John T. Bowen in memory of Eliza M. Bowen; courtesy of Museum of Fine Arts, Boston)

Massachusetts,[1] is signed and dated J. S. Copley 1753. One of Copley's earliest known works, painted when he was only fifteen years old, it depicts a robust young woman staring candidly at the viewer. Seated outdoors in front of a rock outcropping, she rests her left elbow on a classical pedestal and she dangles a string of pearls from her left hand.

Relation of the painting to its source

The painting suffers from being tied too closely to its mezzotint prototype. The composition is an almost exact mirror image of that used in Isaac Beckett's mezzotint after William Wissing's Princess Anne of ca. 1683.[2] Pose, props, and background are all lifted directly from the print. Certain changes, however, were necessary to acclimatize the image to its new American setting. Princess Anne is shown provocatively posed in a landscape setting. Her blouse slips from her shoulders to reveal an enticing amount of bare bosom. Her hair curls lasciviously over her shoulders and a pearl necklace slides suggestively through her fingers as though, having removed the pearls, she will proceed further to disrobe. But Copley reduces the sensual overtones. Mrs. Mann's bodice is decorously raised to ensure sufficient coverage, and

---

[1]Jules David Prown, John Singleton Copley (Cambridge: Harvard University Press, 1966), I, 110.

[2]Charles Coleman Sellers, "Mezzotint Prototypes of Colonial Portraiture: A Survey Based on the Research of Waldon Phoenix Belknap, Jr.," Art Quarterly 20 (1957): 407-68. See especially plate 16.

the alluring gaze of the princess is replaced by a cool stare. However, the suggestive pearls remain intact, producing an oddly discordant note.

First sentence of paragraph is both a transition and a topic sentence: the weakness of the painting

The picture has other problems as well. The young Copley obviously had not yet learned to handle his medium. The brush strokes are long and streaky. The shadows around the nose are a repellent greenish purple and the highlight on the bridge was placed too far to one side. The highlights in the hair were applied while the underlying brown layer was still wet so that instead of gleaming curls he produced dull gray smudges. In addition, textural differentiation is noticeably lacking. The texture of the rock is the same as the skin, which is the same as the satin and the

Concrete details support the paragraph's opening assertion

grass and the pearls. The anatomy is laughable: There is no sense of underlying structure. The arms and neck are the inflated tubes so typical of provincial portraiture. The left earlobe is missing and the little finger on the left hand is disturbingly disjointed. Light too appears to have given Copley trouble. It seems, in general, to fall from the upper left, but the shadows are not consistently applied. And the light-dark contrasts are rather too sharp, probably due to an overreliance on the mezzotint source.

Transition ("Despite its faults") and statement of idea that unifies the paragraph

Despite its faults, however, the painting still represents a remarkable achievement for a boy of fifteen. In the crisp linearity of the design, the sense of weight and bulk of the figure, the hint of a psychological presence, and especially in the rich vibrant color, Copley has already rivaled and even

surpassed the colonial painters of the previous
generation.

Transition ("about seventeen years later") and reassertion of central thesis

In Mrs. Ezekial Goldthwait, about seventeen years later and about four years before Copley went to England, all the early ineptness had disappeared. Copley has arrived at a style that is both uniquely his own and uniquely American; and in this style he achieves a level of quality comparable to any of his English contemporaries.

Brief description of the second picture

The substantial form of Mrs. Goldthwait dominates the canvas. She is seated at a round tilt-top table, one hand extended over a tempting plate of apples, oranges, and pears. A huge column rises in the right-hand corner to fill the void.

Biography and (in rest of paragraph) its relevance to the work

The fifty-seven-year-old Mrs. Goldthwait, wife of a wealthy Boston merchant, was the mother of fourteen children; she was also a gardener noted for her elaborate plantings.[3] Copley uses this fertility theme as a unifying element in his composition. All the forms are plump and heavy, like Mrs. Goldthwait herself. The ripe, succulent fruit, the heavy, rotund mass of the column, the round top of the table--all are suggestive of the fecundity of the sitter.

The most obvious characteristic of the work

The painting is also marked by a painstaking realism. Each detail has been carefully and accurately rendered, from the wart on her forehead to the wood grain of the tabletop to the lustrous gleam of the pearl

---

[3]Prown, 76.

necklace. As a painter of fabrics Copley surpasses all his contemporaries. The sheen of the satin, the rough, crinkly surface of the black lace, the smooth, translucent material of the cuffs--all are exquisitely rendered.

"But" is transitional, taking us from the obvious (clothing) to the less obvious (character)

But the figure is more than a mannequin modeling a delicious dress. She has weight and bulk, which make her physical presence undeniable. Her face radiates intelligence, and her open, friendly personality is suggested by the slight smile at the corner of her lips and by her warm, candid gaze.

Brief reminder of the first work, to clarify our understanding of the second work

The rubbery limbs of Copley's early period have been replaced by a more carefully studied anatomy (not completely convincing, but still a remarkable achievement given that he was unable to dissect or to draw from nude models). There is some sense for the underlying armature of bone and muscle, especially in the forehead and hands. And in her right hand it is even possible to see the veins running under her skin.

Further comparison, again with emphasis on the second work

Light is also treated with far greater sophistication. The chiaroscuro is so strong and rich that it calls to mind Caravaggio's tenebroso. The light falls almost like a spotlight onto the face of Mrs. Goldthwait, drawing her forward from the deep shadows of the background, thereby enhancing the sense of a psychological presence.

Reassertion of the thesis, supported by concrete details

Copley's early promise as a colorist is fulfilled in mature works such as Mrs. Goldthwait. The rich, warm red-brown tones of the satin, the wood, and the column dominate the composition. But the painting is enlivened by a splash of color on either side--on the left by Copley's favorite aqua in the brocade of the chair, and on the right by the red and green punctuation marks of the fruit. The bright white of the cap, set off against the black background, draws attention to the face, while the white of the sleeves performs the same function for the hands.

Summary, but not mere rehash; new details

Color, light, form, and line all work together to produce a pleasing composition. It is pleasing, above all, for the qualities that distinguish it from contemporary English works: for its insistence on fidelity to fact, for its forthright realism, for the lovingly delineated textures, for the crisp clarity of every line, for Mrs. Goldthwait's charming wart and her friendly double chin, for the very materialism that marks this painting as emerging from our pragmatic mercantile society. In these attributes lie the greatness of the American Copleys.

Further summary, again heightening the thesis

Not that I want to say that Copley never produced a decent painting once he arrived in England. He did. But what distinguishes the best of his English works (see, for example Mrs. John Montressor and Mrs. Daniel Denison Rogers)[4] is not the facile, flowery brushwork or the fluttery drapery (which he picked up from current English practice) but the very qualities that also mark the best of his American works--the realism, the sense of personality, the almost touchable textures of the fabrics, and the direct way in

[4]Prown, plates.

which the sitter's gaze engages the viewer. Copley
was a fine, competent painter in England, but it was not
the trip to England that made him great.

[NEW PAGE]

Works Cited

Prown, Jules David. John Singleton Copley. 2 vols.
Cambridge: Harvard University Press, 1966.

Sellers, Charles Coleman. "Mezzotint Prototypes of Colonial
Portraiture: A Survey Based on the Research of Waldon
Phoenix Belknap, Jr." Art Quarterly 20 (1957): 407-68.

✔ Checklist for Writing a Comparison

✔ Is the point of the comparison clear? (Examples: to show that although X and Y superficially resemble each other, they are significantly different; or, to show that X is better than Y; or, to illuminate
X by briefly comparing it to Y)

✔ Are all significant similarities and differences covered?

✔ Is the organization clear? If the chief organizational device is lumping, does the second half of the essay connect closely enough with
the first so that the essay does not divide into two essays? If the chief
organizational device is splitting, does the essay avoid the Ping-Pong
effect? Given the topic and the thesis, is it the best organization?

✔ If a value judgement is offered, is it supported by evidence?

4

How to Write an Effective Essay

The Basic Strategy

- Choose a topic and a tentative thesis
- Generate ideas, for instance by asking yourself questions
- Make a tentative outline of points you plan to make
- Rough out a first draft, working from your outline (don't worry about spelling, punctuation, etc.)
- Make large-scale revisions in your draft by reorganizing, or by adding details to clarify and support assertions, or by deleting or combining paragraphs
- Make small-scale revisions by revising and editing sentences
- Revise your opening and concluding paragraphs
- Have someone read your revised draft and comment on it
- Revise again, taking into account the reader's suggestions
- Read this latest version and make further revisions as needed
- Proofread your final version

All writers must work out their own procedures and rituals, but the following basic suggestions will help you write effective essays. They assume that you take notes on index cards, but you can easily adapt the principles if you use a laptop. If your paper involves using sources, consult also Chapter 8, "Writing a Research Paper."

LOOKING CLOSELY: APPROACHING A FIRST DRAFT

1. **Look at the work or works carefully.**
2. **Choose a worthwhile and compassable subject,** something that interests you and is not so big that your handling of it must be superficial. As you work, shape your topic, narrowing it, for example, from

"Egyptian Sculpture" to "Black Africans in Egyptian Sculpture," or from "Frank Lloyd Wright's Development" to "Wright's Johnson Wax Company as an Anticipation of His Guggenheim Museum."

3. **Keep your purpose in mind.** Although your instructor may ask you, perhaps as a preliminary writing assignment, to jot down your early responses—your initial experience of the work—it is more likely that he or she will ask you to write an analysis in which you will connect details and draw inferences. Almost surely you will be asked to do more than report your responses or to write a description of an object; you probably will be expected to support a *thesis*, that is, to offer an *argument*. Obviously an essay that evaluates a work not only offers a judgment but also supports the judgment with evidence. Yet even a formal analysis presents an argument, holding that the work is constructed in such-and-such a way and that its meaning (or one of its meanings) is communicated by the relationships between the parts.

In thinking about your purpose, remember, too, that your audience will in effect determine the amount of detail that you must give. Although your instructor may in reality be your only reader, probably you should imagine that your audience consists of people pretty much like your classmates—intelligent, but not especially familiar with the topic on which you have recently become a specialist.

4. **Keep looking at the art** you are writing about (or reproductions of it), jotting down notes on all relevant matters.

- You can generate ideas by asking yourself questions such as those given on pages 33–81.
- As you look and think, reflect on your observations and record them.
- When you intend to write about an object in a museum that you are visiting, choose an object that is reproduced on a postcard; the picture will help you to keep the object in mind when you are writing in your room.
- When you have an idea, jot it down; don't assume that you will remember it when you begin writing. A sheet of paper is good for initial jottings, but many people—if they are not taking notes on a word processor—find that it is easiest to use 4″ × 6″ cards.
- Put only one point on each card, and put a brief caption on the card (e.g., Site of *David*); later you can arrange the cards so that the relevant notes are grouped together.

5. When taking notes from secondary sources, do not simply highlight or photocopy.

- Take brief notes, *summarizing* important points and jotting down your own critiques of the material.
- Read the material analytically, thoughtfully, with an open mind and a questioning spirit.
- When you read in this attentive and tentatively skeptical way, you will find that the material is valuable not only for what it tells you but also for the ideas that you yourself produce in responding to it.

Writing your paper does not begin when you sit down to write a draft; rather, it begins when you write your first thoughtful notes.

6. Sort out your note cards, putting together what belongs together. Three separate cards with notes about the texture of the materials of a building, for instance, probably belong together. Reject cards irrelevant to your topic.

7. Organize your packets of cards into a reasonable sequence. Your cards contain ideas (or at least facts that you can think about); now the packets of cards have to be put into a coherent sequence. When you have made a tentative arrangement, review it; you may discover a better way to group your notes, and you may even want to add to them. If so, start reorganizing.

A tripartite organization usually works. For this structure, tentatively plan to devote your opening paragraph(s) to a statement of the topic or problem and a proposal of your hypothesis or thesis. The essay will then take this shape:

- a beginning, in which you identify the work(s) of art that you will discuss, giving the necessary background and, in a sentence or two, setting forth your underlying argument, your thesis
- a middle, in which you develop your argument, chiefly by offering evidence
- a conclusion, in which you wrap things up, perhaps by giving a more general interpretation or by setting your findings in a larger context. (On concluding paragraphs, see pages 146–147.)

In general, organize the material from the simple to the complex in order to ensure intelligibility. If, for instance, you are discussing the composition of a painting, it probably will be best to begin with the most obvious points and then to turn to the subtler but perhaps equally important ones. Similarly, if you are comparing two sculptures, it may be best to move from the most obvious contrasts to the least obvious. When you have arranged your notes into a meaningful sequence of packets, you have approximately divided your material into paragraphs.

8. **Get it down on paper.** Most essayists find it useful to jot down some sort of **outline,** a map indicating the main idea of each paragraph and, under each main idea, supporting details that give it substance. An outline—not necessarily anything highly formal, with capital and lowercase letters and roman and arabic numerals, but merely key phrases in some sort of order—will help you to overcome the paralysis called "writer's block" that commonly afflicts professional as well as student writers. For an example of a student's rough outline, see the jottings on page 92 that were turned into an essay on the sculpture *Seated Statue of Prince Khunera as a Scribe.*

A page of paper with ideas in some sort of sequence, however rough, ought to encourage you that you do have something to say. And so, despite the temptation to sharpen another pencil or to put a new ribbon into the typewriter or to get some new software, the best thing to do at this point is to follow the advice of Isaac Asimov, author of 225 books: "Sit down and start writing."

If you don't feel that you can work from note cards and a rough outline, try another method: Get something down on paper, writing (preferably on a word processor) freely, sloppily, automatically, or whatever, but allow your ideas about what the work means to you and how it conveys its meaning—rough as your ideas may be—to begin to take visible form. If you are like most people, you can't do much precise thinking until you have committed to paper at least a rough sketch of your initial ideas. Later you can push and polish your ideas into shape, perhaps even deleting all of them and starting over, but it's a lot easier to improve your ideas once you see them in front of you than it is to do the job in your head. On paper one word leads to another; in your head one word often blocks another.

Just keep going; you may realize, as you near the end of a sentence, that you no longer believe it. Okay; be glad that your first idea led you to a better one, and pick up your better one and keep going with it. What you are doing is, in a sense, by trial and error pushing your way not only toward clear expression but also toward sharper ideas and richer responses.

REVISING: ACHIEVING A READABLE DRAFT

Good writing is *rewriting*. The evidence? Heavily annotated drafts by Chekhov, Hemingway, Tolstoy, Yeats, Woolf—almost any writer you can name. Of course it is easy enough to spill out words, but, as the dramatist Richard Sheridan said two hundred years ago, "Easy writing's curst hard reading." In the words of the German novelist Thomas Mann, "A writer is someone for whom writing is more difficult than it is for other people." It is difficult because writers care; they care about making sense, so they will take time to find the exact word, the word that enables them to say precisely what they mean, so their readers will get it right. And they care about holding a reader's attention.

1. **Keep looking and thinking,** asking yourself questions and providing tentative answers, searching for additional material that strengthens or weakens your main point, and take account of it in your outline or draft.

Now is probably the time to think about a title for your essay. It is usually a good idea to let your reader know what your topic is—which works of art you will discuss—and what your approach is, for instance, "A Formal Analysis of *Prince Khunera as a Scribe*," or "Van Gogh's *Self-Portrait as a Priest:* A Psychoanalytic Approach." At this stage your title is still tentative, but thinking about a title will help you to organize your thoughts and to determine which of your notes are relevant and which are not. Remember, the title that you settle on is the first part of the paper that your reader encounters. You will gain the reader's goodwill by providing a helpful, interesting title.

2. **With your outline or draft in front of you, write a more lucid version,** checking your notes for fuller details. If, as you work, you find that some of the points in your earlier jottings are no longer relevant, eliminate them; but make sure that the argument flows from one point to the next. It is not enough to keep your thesis in mind; you must keep it in the reader's mind. As you write, your ideas will doubtless become clearer, and some may prove to be poor ideas. (We rarely know exactly what our ideas are until we set them down on paper. As the little girl said, replying to the suggestion that she should think before she spoke, "How do I know what I think until I see what I say?") Not until you have written a draft do you really have a strong sense of what you feel and know, and of how good your essay may be.

If you have not already made an outline at this stage, it is probably advisable to make one, to ensure that your draft is reasonably organized.

✍ A RULE FOR WRITERS:

Put yourself in the reader's shoes to make sure that the paper not only has an organization but that the organization will be clear to your reader. If you imagine yourself as the reader of the draft, you may find that you need to add transitions, clarify definitions, and provide additional supporting evidence.

Jot down, in sequence, each major point and each subpoint. You may find that some points need amplification, or that a point made on page 3 really ought to go on page 1.

3. **After a suitable interval, preferably a few days, again revise and edit the draft.** To write a good essay you must be a good reader of the essay you are writing. We're not talking at this stage about proofreading or correcting spelling errors, though you will need to do that later. Van Gogh said, "One becomes a painter by painting." Similarly, one becomes a writer by writing—and by rewriting, or revising. In revising their work, writers ask themselves such questions as

- Do I mean what I say?
- Do I say what I mean? (Answering this question will cause you to ask yourself such questions as, Do I need to define my terms? add examples to clarify? reorganize the material so that a reader can grasp it?)

During this part of the process of writing, you want to read the draft in a skeptical frame of mind. In taking account of your doubts, you will probably unify, organize, clarify, and polish the draft.

- **Unity** is achieved partly by eliminating irrelevancies. These may be small (a sentence or two) or large (a paragraph or even a page or two). You wrote the material and you are fond of it, but if it is irrelevant you must delete it.
- **Organization** is largely a matter of arranging material into a sequence that will assist the reader to grasp the point. If you reread your draft and jot down a paragraph outline—a series of sentences, one under the other, each sentence summarizing one paragraph—you can then see if the draft has a reasonable

organization, a structure that will let the reader move easily from the beginning to the end.

- **Clarity** is achieved largely by providing concrete details, examples, and quotations to support generalizations, and by providing helpful transitions ("for instance," "furthermore," "on the other hand," "however").
- **Polish** is small-scale revision. One deletes unnecessary repetitions, combines choppy sentences into longer sentences, and breaks overly long sentences into shorter sentences.

If you have written your draft on a word processor, do *not* try to revise it on the monitor. Print the entire draft, and then read it—as your reader will be reading it—page by page, not screen by screen. Almost surely you will detect errors in a hard copy that you miss on the screen. Only by reading the printed copy will you be able to see if, for instance, paragraphs are too long.

After producing a draft that seems good enough to show to someone, writers engage in yet another activity. They edit. **Editing** includes such work as checking the accuracy of quotations by comparing them with the original, checking a dictionary for the spelling of doubtful words, and checking a handbook for doubtful punctuation—for instance, whether a comma or a semicolon is needed in a particular sentence.

PEER REVIEW

Your instructor may encourage (or even require) you to discuss your draft with another student or with a small group of students. That is, you may be asked to get a review from your peers. Such a procedure is helpful in several ways. First, it gives the writer a real audience, readers who can point to what pleases or puzzles them, who make suggestions, who may often disagree (with the writer or with each other), and who frequently, though not intentionally, *misread.* Though writers don't necessarily like everything they hear (they seldom hear "This is perfect. Don't change a word!"), reading and discussing their work with others almost always gives them a fresh perspective on their work, and a fresh perspective may stimulate thoughtful revision. (Having your intentions *misread,* because your writing isn't clear enough, can be particularly stimulating.)

✔ Checklist for Peer Review

Read each draft once, quickly. Then read it again and jot down brief responses to the following questions.

✔ What is the essay's topic? Is it one of the assigned topics, or a variation of one of them? Is the title appropriate? Does the draft show promise of fulfilling the assignment?

✔ Looking at the essay as a whole, what thesis (main idea) is stated or implied? If implied, try to state it in your own words.

✔ Is the thesis plausible? How might it be strengthened?

✔ Looking at each paragraph separately:

  ✔ What is the basic point?

  ✔ How does each paragraph relate to the essay's main idea or to the previous paragraph?

  ✔ Should some paragraphs be deleted? be divided into two or more paragraphs? be combined? be put elsewhere? (If you outline the essay by jotting down the gist of each paragraph, you will get help in answering these questions.)

  ✔ Is each sentence clearly related to the sentence that precedes and to the sentence that follows?

  ✔ Is each paragraph adequately developed? Are there sufficient details to support the generalizations?

  ✔ Are the introductory and concluding paragraphs effective?

✔ Are the necessary illustrations included, and are they adequately identified?

✔ What are the paper's chief strengths?

✔ Make at least two specific suggestions that you think will assist the author to improve the paper.

The writer whose work is being reviewed is not the sole beneficiary. When students regularly serve as readers for each other, they become better readers of their own work and consequently better revisers. And, as you probably know, learning to write is in large measure learning to read.

If peer review is a part of the writing process in your course, the instructor may distribute a sheet with suggestions and questions. The preceding checklist is an example of such a sheet.

✍ **A RULE FOR WRITERS**
(ATTRIBUTED TO TRUMAN CAPOTE):

Good writing is rewriting.

PREPARING THE FINAL VERSION

1. **If you have received comments from a reader, consider them carefully.** Even if you disagree with them, they may alert you to places in your essay that need revision, such as clarification.

In addition, if a friend, a classmate, or another peer reviewer has given you some help, acknowledge that help in a footnote or endnote. (If you look at almost any book or any article in *The Art Bulletin* you will notice that the author acknowledges the help of friends and colleagues. In your own writing follow this practice.) Here are sample acknowledgments from papers by students:

> I wish to thank Anna Aaron for numerous valuable suggestions.

> I wish to thank Paul Gottsegen for calling my attention to passages that needed clarification, and Jane Leslie for suggesting the comparison with Orozco's murals at Dartmouth College.

> Emily Andrews called my attention to recent studies of Mayan art.

> I am indebted to Louise Cort for explaining how Shigaraki ceramics were built and fired.

2. **Write, type, or print a clean copy,** following the principles concerning margins, pagination, footnotes, and so on, set forth in Chapter 9. If you have borrowed any ideas, be sure to give credit, usually in footnotes, to your sources. Remember that plagiarism is not limited to the unacknowledged borrowing of words; a borrowed idea, even when put into your own words, requires acknowledgment. (On giving credit to sources, see pages 238–62.)

3. **Proofread and make corrections** as explained on page 264.

In short, ask these questions:

- Is the writing true (do you have a point that you state accurately)?
- Is the writing good (do your words and your organization clearly and effectively convey your meaning)?

All of this adds up to Mrs. Beeton's famous recipe: "First catch your hare, then cook it."

5

Style in Writing

PRINCIPLES OF STYLE

Writing is hard work (Lewis Carroll's school in *Alice's Adventures in Wonderland* taught reeling and writhing), and there is no point in fooling ourselves into believing that it is all a matter of inspiration. Many of the books that seem, as we read them, to flow so effortlessly were in fact the product of innumerable revisions. "Hard labor for life" was Joseph Conrad's view of his career as a writer. This labor, for the most part, is not directed to prettifying language but to improving one's thoughts and then getting the words that communicate these thoughts exactly. There is no guarantee that effort will pay off, but failure to expend effort is sure to result in writing that will strike the reader as confused. It won't do to comfort yourself with the thought that you have been misunderstood. You may know what you *meant to say*, but your reader is the judge of what indeed you *have said*. Keep in mind Henri Matisse's remark: "When my words were garbled by critics or colleagues, I considered it my fault, not theirs, because I had not been clear enough to be comprehended."

Many books have been written on the elements of good writing, but the best way to learn to write is to do your best, show it to a friend, think about the response and revise accordingly, revise it a few days later, hand it in, and then study the annotations an experienced reader puts on your essay. In revising the annotated passages, you will learn what your weaknesses are in writing. After drafting your next essay, put it aside for a day or so, then reread it, preferably aloud. You may find much that bothers you. (If you read it aloud, you will probably catch choppy sentences, needless repetitions, and unpleasant combinations of words, such as "We see in the sea . . . .") If the argument does not flow, check to see whether your organization is reasonable and whether you have made adequate transitions. Do not hesitate to delete interesting but irrelevant material that obscures the argument. Make the necessary revisions again and again if there is time. Revision is indispensable if you wish to avoid (in Somerset Maugham's words) "the impression of writing with the stub of a blunt pencil."

Even though the best way to learn to write is by writing and by heeding the comments of your readers, a few principles can be briefly set forth here. These principles will not suppress your particular voice; rather, they will get rid of static, enabling your voice to come through effectively. You have something to say, but you can say it only after your throat is cleared of "Well, what I meant was" and "It's sort of, well, you know." Your readers do *not* know; they are reading in order to know. This chapter will help you let your individuality speak clearly.

GET THE RIGHT WORD

Denotation

Be sure the word you choose has the right explicit meaning, or denotation. Don't say "tragic" when you mean "pathetic," "carving" when you mean "modeling," or "print" when you mean a photographic reproduction of a painting.

Connotation

Be sure the word you choose has the right association or implication—that is, the right connotation. Here is an example of a word with the wrong connotation for its context: "Close study will *expose* the strength of Klee's style." "Reveal" would be better than "expose" here; "expose" suggests that some weakness will be brought to light, as in "Close study will expose the flimsiness of the composition."

Sometimes our prejudices blind us to the unfavorable connotations of our vocabulary. Writing about African architecture, one student spoke of "mud huts" throughout the paper; a more respectful term for the same type of building would be "clay house" or "earthen compound." When you submit your paper to a colleague for peer review, urge your reviewer to question your use of words that may have inappropriate connotations.

Concreteness

Catch the richness, complexity, and uniqueness of what you see. Do not write "His expression lacks emotion" if you really mean the expression is icy or indifferent. But concreteness is not only a matter of getting the exact word—no easy job in itself. If your reader is (so to speak) to see what you are getting at, you have to provide some details. Instead of writing "The influence of photography on *X* is small," write "The influence of photography is evident in only six paintings."

Compare the rather boring statement, "Thirteenth-century sculpture was colored," or even "Thirteenth-century sculpture was brightly painted and sometimes adorned with colored glass," with these sentences rich in detail:

Concrete

> Color was an integral part of sculpture and its setting. Face and hands were given their natural colors; mouth, nose, and ears were slightly emphasized; the hair was gilded. Dresses were either covered in flowers or painted in vigorous colors; ornaments, buckles, and hems were highlighted by brilliant colors or even studded with polished stones or colored glass. The whole portal looked like a page from an illuminated manuscript, enlarged on a vast scale.
>
> —Marcel Aubert, *The Art of the High Gothic Era,*
> trans. Peter Gorge (New York: Crown, 1965), 60

A note on the use of "this" without a concrete reference. Avoid using "this" when you mean "what I have been saying." Your reader does not know if "this" refers to the preceding clause, sentence, paragraph, or page.

Imprecise

> She did not begin to paint until she was fifty. Moreover, she did not try to sell her work until at least ten years later. This proved to have advantages.

To what does "this" refer? That she did not try to sell her work for ten years? That she did not paint until she was fifty? Both? Perhaps something even earlier in the paragraph? It turned out that "this" refers to the points made in the first two sentences quoted, and the student successfully revised the third sentence by providing specific references for "this":

Precise and clear because "this" is made specific

> She did not begin to paint until she was fifty. Moreover, she did not try to sell her work until at least ten years later. This late start and lack of concern for the market proved to have advantages.

A Note on Technical Language

Discussions of art, like, say, discussions of law, medicine, the dance, and for that matter, cooking and baseball, have given rise to technical terminology. A cookbook will tell you to blend, boil, or bake, and it will speak of a "slow" oven (300 degrees), a "moderate" oven (350 degrees), or a "hot" oven (400 degrees). These are technical terms in the world of cookery, and no one objects that it is preposterous to define a hot oven as 400 degrees when everyone knows that a hot day is 90 degrees.

In watching a baseball game we find ourselves saying, "I think the hit-and-run is on," or "He'll probably bunt." We use these terms because they convey a good deal in a few words; like "a hot oven," they are clear and precise. Technical language is illuminating—provided (1) it is used accurately, and (2) the audience is familiar with the language. How can an audience not be familiar with language? New ways of thinking—new systems of thought, such as Structuralism and Queer Theory—produce new language, language that is meaningful to the initiated but puzzling to others. If it is used imprecisely, or if it is used unnecessarily, in an effort to impress the hearer, it becomes jargon to everybody.

In writing about art you will, for the most part, use the same language that you use in other courses. You will not needlessly introduce unusual words, but you *will* use the language of art history when it enables you to be clear, concise, and accurate. Some specialized words are known to most native speakers ("etching," "perspective," "still life"); some are known chiefly to highly educated people ("bas relief," "gouache," "vellum"), and some are known only to people who have read a fair amount of critical theory concerning art or literature ("episteme," "poststructuralism," "semiotics"). Obviously readers who for the first time encounter the word "episteme" are aware that they are in new territory, and they either try to understand the word from the context, or they consult a reference work. But another sort of technical word is more dangerous, a word like "hot," which means one thing when talking about the weather and another thing when talking about using an oven. For instance, among the words that look innocent but that recently have acquired highly technical meanings in discussions of art are "appropriation," "code," "discourse," "erasure," and "fetish." How technical? "Fetish" gets a two-and-a-half-page discussion in *Artwords* (1997), a dictionary of current critical terms by Thomas Patin and Jennifer McLerran, and it gets eleven pages in *Critical Terms for Art History* (1996), ed. Robert S. Nelson and Richard Shiff. If you encounter these words in a book or in a lecture, and if you think that they are impressive ways of saying something that otherwise sounds commonplace, do *not* use them in your own writing. After all, in an essay you wouldn't speak of a house as a "domicile," or a professor as a "pedagogue." On the other hand, if indeed you understand the specialized use of certain terms, and if they strike you as the best way to make your point, of course you can use them—though if you are writing for a general public you will have to clarify them.

Tone

Remember, when you are writing, *you* are the teacher. You are trying to help someone to see things as you see them, and it is unlikely that either solemnity or heartiness will help anyone see anything your way. There is rarely a need to write that Daumier was "incarcerated" or (at the other extreme) "thrown into the clink." "Imprisoned" or "put into prison" will probably do the job best. Be sure, also, to avoid shifts in tone. Consider this passage, from a book on modern sculpture:

Faulty

> We forget how tough it was to make a living as a sculptor in this period. Rare were supportive critics, dealers, and patrons.

Although "tough" is pretty casual ("difficult" might be better), "tough" probably would have been acceptable if it had not been followed, grotesquely, by the pomposity of "Rare were supportive critics." The unusual word order (versus the normal order, "Supportive critics were rare") shows a straining for high seriousness that is incompatible with "tough."

Nor will it do to "finagle" with an inappropriate expression by putting it in "quotes." As the previous sentence indicates, the apologetic quotation marks do not make such expressions acceptable, only more obvious and more offensive. The quotation marks tell the reader that the writer knows he or she is using the wrong word but is unwilling to find the right word. If for some reason a relatively low word is the right one, use it and don't apologize with quotation marks.

✍ A RULE FOR WRITERS:

The words that a writer puts on the page will convey to the reader an image of the writer; good writers take pains to make sure that the image is favorable.

The lesson? As Buffon said two hundred years ago, "The style is the man," to which we can add "or the woman." And, as E. B. White said in our own generation, "No writer long remains incognito."

Repetition

Although some repetitions—say, of words or phrases like "surely" and "it is noteworthy that"—reveal a tic that ought to be cured by revision, don't be afraid to repeat a word if it is the best word. Notice that in the following paragraph the writer does not hesitate to repeat "Impressionism," "Impressionist," "face," and "portrait" (three times each) and "portraiture" and "photography" (twice each).

Effective

> We can follow the decline of portraiture within Impressionism, the art to which van Gogh assumed allegiance. The Impressionist vision of the world could hardly allow the portrait to survive; the human face was subjected to the same evanescent play of color as the sky and sea; for the eyes of the Impressionist it became increasingly a phenomenon of surface, with little or no interior life, at most a charming appearance vested in the quality of a smile or a carefree glance. As the Impressionist painter knew only the passing moment in nature, so he knew only the momentary face, without past or future; and of all its moments, he preferred the most passive and unconcerned, without trace of will or strain, the outdoor, summer holiday face. Modern writers have supposed that it was photography that killed portraiture, as it killed all realism. This view ignores the fact that Impressionism was passionately concerned with appearances, and was far more advanced than contemporary photography in catching precisely the elusive qualities of the visible world. If the portrait declines under Impressionism it is not because of the challenge of the photographer, but because of a new conception of the human being. Painted at this time, the portraits of van Gogh are an unexpected revelation. They are even more surprising if we remember that they were produced just as his drawing and color was becoming freer and more abstract, more independent of nature.
>
> —Meyer Schapiro, *Vincent van Gogh*
> (New York: Abrams, 1952), 16–17

When you repeat words or phrases, or when you provide clear substitutes (such as "he" for "van Gogh"), you are helping the reader to keep step with your developing thoughts.

An ungrounded fear of repetition often produces a vice known as *elegant variation*. Having mentioned "painters," the writer then speaks of "artists," and then (more desperately) of "men and women of the brush." This use of synonyms is far worse than repetition; it strikes the reader as silly. Or it may be worse than silly. Consider:

Confusing

> Corot attracted the timid painters of his generation; bolder artists were attracted to Manet.

The shift from "painters" to "artists" makes us wonder if perhaps Manet's followers—but not Corot's—included etchers, sculptors, and others. Probably the writer did *not* mean any such thing, but the words prompt us to think in the wrong direction.

Be especially careful not to use variations for important critical terms. If, for instance, you are talking about "nonobjective art," don't switch to "abstract art" or "nonrepresentational art" unless you tell the reader why you are switching.

The Sound of Sense, the Sense of Sound

Avoid jingles and other repetitions of sound, as in these examples:

Annoying

> The reason the season is autumn . . .
>
> Circe certainly . . .
>
> Michelangelo's Medici monument . . .

These irrelevant echoes call undue attention to the words and thus get in the way of the points you are making. But wordplay can be effective when it contributes to meaning. For example, in this sentence:

Effective

> The walls of Sian both defended and defined the city.

The echo of "defended" in "defined" nicely emphasizes the unity in this duality.

WRITE EFFECTIVE SENTENCES

Economy

Say everything relevant, but say it in the fewest words possible. Consider the following sentence:

Wordy

> There are a few vague parts in the picture that give it a mysterious quality.

This sentence can be written more economically:

Revised

> A few vaguely defined parts give the picture a mysterious quality.

Nothing has been lost by the deletion of "There are" and "that." An even more economical version could be worded:

Further Revised

> A few vague parts add mystery to the picture.

The original version says nothing that the second version does not say, and it says nothing that the third version—nine words against fifteen—does not say. If you find the right nouns and verbs, you can often delete adjectives and adverbs. (Compare "a mysterious quality" with "mystery.") Something is wrong with a sentence when you can delete words and not sense the loss. A chapter in a recent book begins:

Wordy

> One of the principal and most persistent sources of error that tends to bedevil a considerable proportion of contemporary analysis is the assumption that the artist's creative process is a wholly conscious and purposive type of activity.

Well, there is something of interest here, but it comes along with a lot of hot air. Why that weaseling ("*tends to* bedevil," "*a considerable* proportion"), and why "type of activity" instead of "activity"? Those spluttering *p*'s ("principal and most persistent," "proportion," "process," "purposive") are a giveaway: The writer has not sufficiently revised his writing. It is not enough to have an interesting idea; the job of writing requires *re*writing, revising. The writer of this passage should have revised it—perhaps on rereading it an hour or a day later—and produced something like this:

Revised

> One of the chief errors bedeviling much contemporary criticism is the assumption that the artist's creative process is wholly conscious and purposive.

Possibly the author thought that a briefer and clearer statement would not do justice to the complexity of the main idea, but more likely he simply neglected to reread and revise. The revision says everything that the original says, only better.

When we are drafting an essay we sometimes put down what is more than enough. There's nothing wrong with that—anything goes in a draft—but when we revise we need to delete the redundancies. Here is part of a descriptive entry from an exhibition catalog:

Redundant

> A big black bird with a curved beak is perched on a bare, wintry branch that has lost all its leaves.

If the branch is "bare," it has lost its leaves. No need to write it twice.

The **passive voice** (wherein the subject receives the action) is a common source of wordiness. In general, do not write "The sculpture was carved by Michelangelo" (the subject—"the sculpture"—receives the action—"was carved"). Instead, use the **active voice,** in which the subject acts on the object: "Michelangelo carved the sculpture" (the subject—"Michelangelo"—acts—"carved the sculpture"). Even though the revision is a third shorter, it says everything that the longer version says. Often the passive is needlessly vague, as in "It is believed that . . ." when the writer means "Most people believe" or "Most art historians believe . . . ."

Consider the following passage, the opening paragraph of an essay on the classical aspects of a library at a women's college:

> A person walking by Jackson Library (1908-13) is struck by its classical design. The symmetry of the façade is established by the regularly spaced columns of the Ionic order on the first story of the building, and by pilasters on the second level. In the center of the lower tier are two bronze doors: On the left door a relief is seen, depicting

Sapientia (Wisdom), and on the right is seen the image of Caritas (Charity). The Greco-Roman tradition is furthered by the two bronze statues on either side of the entrance. On the left is Vesta (goddess of the hearth) and on the right Minerva (goddess of wisdom). Through the use of classical architecture and Greco-Roman images, an image is conveyed--one which Charleston College hopes to create in its women.

Although, the paragraph is richly informative, it is sluggish, chiefly because the writer keeps using the passive voice: *A person . . . is struck by; symmetry . . . is established by; a relief is seen; on the right is seen; tradition is furthered by; an image is conveyed.*

Converting some or all of these expressions into the active voice will greatly improve the passage. (A second weakness, however, is the monotony of the sentence structure—subject, verb, object. Notice that in the revision, sentences are more varied. Many versions are possible; try your hand at producing one.)

Revised

Walking by Jackson Library (1908-13), one notices the classical design. Regularly spaced columns of the Ionic order on the first story, and pilasters on the second, establish the symmetry of the façade. In the center of the lower tier are two bronze doors: On the left door a relief depicts Sapientia (Wisdom), and on the right Caritas (Charity). A bronze statue on each side of the entrance (Vesta, goddess of the hearth, on the left, and Minerva, goddess of wisdom, on the right) furthers the Greco-Roman tradition. Charleston College hopes, through the use of classical architecture and Greco-Roman sculpture, to inspire in its women particular ideals.

Yet the passive voice has its uses, for instance, when the doer is unknown ("The picture was stolen Monday morning") or is unimportant ("Drawings should be stored in light-proof boxes) or is too obvious to be mentioned ("The inscription has never been deciphered").

In short, use the active voice rather than the passive voice unless you believe that the passive especially suits your purpose.

Parallels

Use parallels to clarify relationships. Few of us are likely to compose such deathless parallels as "I came, I saw, I conquered," or "of the people, by the people, for the people," but we can see to it that coordinate expressions correspond in their grammatical form. A parallel such as "He liked to draw and to paint" (instead of "He liked drawing and to paint") neatly says what the writer means. Notice the clarity and the power of the following sentence on Henri de Toulouse-Lautrec:

Parallel

> The works in which he records what he saw and understood contain no hint of comment—no pity, no sentiment, no blame, no innuendo.
> —Peter and Linda Murray, *A Dictionary of Art and Artists,* 4th ed.
> (Harmondsworth, England: Penguin, 1976), 451

The following wretched sentence seems to imply that "people" and "California and Florida" can be coordinate:

Faulty

> The sedentary Pueblo people of the southwestern states of Arizona and New Mexico were not as severely affected by early Spanish occupation as were California and Florida.

This sort of fuzzy writing is acceptable in a first or even a second draft, but not in a finished essay.

Subordination

A word about short sentences: They can, of course, be effective ("Rembrandt died a poor man"), but unless what is being said is especially weighty, short sentences seem childish. They may seem choppy, too, because the periods keep slowing the reader down. Consider these sentences:

Choppy

> He was assured of government support. He then started to dissociate himself from any political aim. A long struggle with the public began.

There are three ideas here, but they are not worth three separate sentences. The choppiness can be reduced by combining them, subordinating some parts to others. In **subordinating,** make sure that the less important element is subordinate to the more important. In the following example the first clause ("As soon as he was assured of government support"), summarizing the writer's previous sentences, is a subordinate or dependent clause; the new material is made emphatic by being put into two independent clauses:

Revised

> As soon as he was assured of government support, he started to dissociate himself from any political aim, and the long struggle with the public began.

The second and third clauses in this sentence, linked by "and," are coordinate—that is, of equal importance.

We have already discussed parallels ("I came, I saw, I conquered") and pointed out that parallel or coordinate elements should appear so in the sentence. The following line gives van Gogh and his brother Theo equal treatment:

> Van Gogh painted at Arles, and his brother Theo supported him.

This is a **compound sentence**—composed of two or more clauses that can stand as independent sentences but that are connected with a coordinating conjunction such as *and, but, for, nor, yet;* or with a correlative conjunction such as *not only . . . but also;* or with a conjunctive adverb such as *also* or *however* (these require a semicolon); or with a colon, semicolon, or (rarely) a comma.

A **complex sentence** (an independent clause and one or more subordinate clauses), however, does not give equal treatment to each clause; whatever is outside the independent clause is subordinate, less important. Consider this sentence:

> Supported by Theo's money, van Gogh painted at Arles.

The writer puts van Gogh in the independent clause ("van Gogh painted at Arles"), subordinating the relatively unimportant Theo. Notice, by the way, that emphasis by subordination often works along with emphasis by position. Here the independent clause comes after the subordinate clause; the writer appropriately puts the more important material at the end—that is, in the more emphatic position.

Had the writer wished to give Theo more prominence, the passage might have run:

> Theo provided money, and van Gogh painted at Arles.

Here Theo stands in an independent clause, linked to the next clause by "and." Each of the two clauses is independent, and the two men (each in an independent clause) are now of approximately equal importance.

If the writer wanted instead to deemphasize van Gogh and to emphasize Theo, the sentence might read:

While van Gogh painted at Arles, Theo provided the money.

Here van Gogh is reduced to the subordinate clause ("while van Gogh painted at Arles") and Theo is given the dignity of the only independent clause ("Theo provided the money"). (Notice again that the important point is also in the emphatic position, near the end of the sentence. A sentence is likely to sprawl if an independent clause comes first, preceding a long subordinate clause of lesser importance, such as the sentence you are now reading.)

In short, though simple sentences and compound sentences have their place, they make everything of equal importance. Since everything is not of equal importance, you must often write complex and compound-complex sentences, subordinating some ideas to other ideas.

But note: You need not worry about subtle matters of emphasis while you are drafting your essay. When you reread the draft, however, you may feel that certain sentences dilute your point, and it is at this stage that you should check to see if you have adequately emphasized what is important.

WRITE UNIFIED AND COHERENT PARAGRAPHS

A paragraph is normally a group of related sentences. These sentences explore one idea in a coherent (organized) way.

Unity

If your essay is some five hundred words long—about two double-spaced typewritten pages—you probably will not break it down into more than four or five parts or paragraphs. (But you *should* break your essay down into paragraphs—that is, into coherent blocks that give the reader a rest between them. One page of typing is about as long as you can go before the reader needs a slight break.) A paper of five hundred words with a dozen paragraphs is probably faulty not because it has too many ideas but because it has too few *developed* ideas. A short paragraph—especially one consisting of a single sentence—is usually anemic; such a paragraph may be acceptable when it summarizes a highly detailed previous paragraph or group of paragraphs, or when it serves as a transition between two complicated paragraphs, but usually summaries and transitions can begin the next paragraph.

The unifying idea in a paragraph may be explicitly stated in a **topic sentence.** Most commonly, the topic sentence is the first sentence, forecasting what is to come in the rest of the paragraph; or it may be the second sentence, following a transitional sentence. Less commonly, it is the last sentence, summarizing the points that the paragraph's earlier sentences have made. Least commonly—but thoroughly acceptable—the paragraph may have no topic sentence, in which case it has a **topic idea**—an idea that holds the sentences together although it has not been explicitly stated. Whether explicit or implicit, an idea must unite the sentences of the paragraph. If your paragraph has only one or two sentences, the chances are that you have not adequately developed its idea.

A paragraph can make several points, but the points must be related, and the nature of the relationship must be indicated so that there is, in effect, a single unifying point. Here is a satisfactory paragraph about the first examples of Egyptian sculpture in the round. (The figures to which the author refers are not reproduced here.)

Unified

> Sculpture in the round began with small, crude human figures of mud, clay, and ivory (Fig. 4). The faces are pinched out of the clay until they have a form like the beak of a bird. Arms and legs are long rolls attached to the slender bodies of men, while the hips of the women's figures are enormously exaggerated. A greater variety of attitudes and better workmanship are found in the ivory figurines which sometimes have the eye indicated by the insertion of a bead (Fig. 4). It is the carving of animals, however, such as the ivory hippopotamus from Mesaeed in Fig. 4, or the pottery figure (Fig. 6) which points the way toward the rapid advance which was to be made in the Hierakonpolis ivories and in the small carvings of Dynasty I.
>
> —William Stevenson Smith. *Ancient Egypt*, 4th ed.
> (Boston: Museum of Fine Arts, 1960), 20

Smith is talking about several kinds of objects, but his paragraph is held together by a unifying topic idea. The idea is this:

> Although most of the early sculpture in the round is crude, some pieces anticipate the later, more skilled work.

Notice, by the way, that Smith builds his material to a climax, beginning with the weakest pieces (the human figures) and moving to the best pieces (the animals).

The beginning and especially the end of a paragraph are usually the most emphatic parts. A beginning may offer a generalization that the rest of the paragraph supports. Or the early part may offer details, preparing for the generalization in the later part. Or the paragraph may move from cause to effect. Although no rule can cover all paragraphs (except that all must make a point in an orderly way), one can hardly go wrong in making the first sentence either a transition from the previous paragraph or a statement of the paragraph's topic. Here is a sentence that makes a transition and states the topic:

> Not only representational paintings but also abstract paintings have a kind of subject matter.

This sentence gets the reader from subject matter in representational paintings (which the writer has been talking about) to subject matter in abstract paintings (which the writer goes on to talk about).

Consider the following two effective paragraphs on Anthony Van Dyck's portrait of Charles I (see page 140).

Unified

> Rather than begin with an analysis of Van Dyck's finished painting of Charles I, let us consider the problem of representation as it might have been posed in the artist's mind. Charles I saw himself as a cavalier or perfect gentleman, a patron of the arts as well as the embodiment of the state's power and king by divine right. He prided himself more on his dress than on robust and bloody physical feats. Van Dyck had available to him precedents for depicting the ruler on horseback or in the midst of a strenuous hunt, but he set these aside. How then could he show the regal qualities and sportsmanship of a dismounted monarch in a landscape? Compounding the artist's problem was the King's short stature, just about 5 feet, 5 inches. To have placed him next to his horse, scaled accurately, could have presented an ungainly problem of their relative heights. Van Dyck found a solution to this last problem in a painting by Titian, in which a horse stood with neck bowed, a natural gesture that in the presence of the King would have appropriate connotations. Placing the royal pages behind the horse and farther from

Anthony Van Dyck, *Charles I*, ca. 1635. Oil on canvas, 8′1″ × 6′11½″. (The Louvre, Paris; Art Resource, NY/Lauros-Giraudon)

the viewer than the King reduced their height and obtrusiveness, yet furnished some evidence of the ruler's authority over men. Nature also is made to support and suitably frame the King. Van Dyck stations the monarch on a small rise and paints branches of a tree overhead to resemble a royal canopy. The low horizon line and our point of view, which allows the King to look down on us, subtly increase the King's stature. The restful stance yet inaccessibility of Charles depends largely upon his pose, which is itself a work of art, derived from art, notably that of Rubens. Its casualness is deceptive; while seemingly at rest in an informal moment, the King is every inch the perfect gentleman and chief of state. The cane was a royal prerogative in European courts of the time, and its presence along with the sword symbolized the gentleman-king.

Just as the subtle pose depicts majesty, Van Dyck's color, with its regal silver and gold, does much to impart grandeur to the painting and to achieve a sophisticated focus on the King. The red, silver, gold, and black of his costume are the most saturate and intense of the painting's colors and contrast with the darker or less intense coloring of adjacent areas. Largely from Rubens, Van Dyck had learned the painterly tricks by which materials and textures could be vividly simulated, so that the eye moves with pleasure from the silvery silken sheen of the coat to the golden leather sword harness and then on to the coarser surface of the horse, with a similar but darker combination of colors in its coat and mane. Van Dyck's portrait is evidence that, whatever one's sympathy for the message, the artist's virtuosity and aesthetic can still be enjoyed.

—Albert E. Elsen, *Purposes of Art,* 3rd ed.
(New York: Holt, 1972), 221–22

Let's pause to look at the structure of these two paragraphs. The first begins in effect by posing a question (What were the problems Van Dyck faced?) and then goes on to discuss Van Dyck's solutions. This paragraph could have been divided into two, the second beginning "Nature also"; that is, if Elsen had felt that the reader needed a break he could have provided it after the discussion of the king's position and before the discussion of nature and the king's pose within nature. But in fact a reader can take in all of the material without a break, and so the topic idea is "Van Dyck's solutions to the problem."

The second paragraph grows nicely out of the first, largely because Elsen begins the second paragraph with a helpful transition, "Just as." The topic idea here is the relevance of the picture's appeal through color; the argument is supported with concrete details, and the paragraph ends by pushing the point a bit further: The colors in the painting not only are relevant to the character portrayed but also have a hold on us.

Coherence

When a paragraph has not only unity but also a structure, then it has coherence; its parts fit together. If you make sure that each sentence is properly related to the preceding and the following sentences, your writing will flow nicely. Nothing is wrong with such obvious **transitions** as *moreover, furthermore, in addition* (these transitions indicate amplification); *but, on the contrary, however, nevertheless, although* (these transitions indicate contrast or concession); *in short, briefly, in other words* (restatement); *finally, therefore, to sum up* (conclusion). But—a transition that tells you to expect some sort of change of direction—transitions should not start every sentence (they can be buried thus: "Degas, moreover, . . ."), and transitions need not appear anywhere in the sentence.

The point is not that transitions must be explicit but that the argument must proceed clearly. The gist of a paragraph might run thus: "Speaking broadly, there were in Japan two traditions of portraiture. . . . The first . . . . The second . . . . The chief difference . . . . But both traditions . . . ."

Consider the following lucid paragraph from an essay on Auguste Rodin's *Walking Man:*

Coherent

> *L'Homme qui marche* is not really walking. He is staking his claim on as much ground as his great wheeling stride will encompass. Though his body's axis leans forward, his rearward left heel refuses to lift. In fact, to hold both feet down on the ground, Rodin made the left thigh (measured from groin to patella) one-fifth as long again as its right counterpart.
>
> —Leo Steinberg, *Other Criteria* (New York: Oxford, 1972), 349

Notice how easily we move through the paragraph: The figure "is not . . . . He is . . . . Though . . . . In fact . . . ." These little words take us neatly from point to point.

Introductory Paragraphs

Vasari, in *Lives of the Painters* (1560, 1568), tells us that Fra Angelico "would never take his pencil in his hand till he had first uttered a prayer." One can easily understand his hope for divine assistance. Beginning a long poem, Lord Byron aptly wrote, "Nothing so difficult as a beginning." Of course, your job is made easier if your instructor has told you to begin your analysis with some basic facts: identification of the object (title, museum, museum number), subject matter (e.g., mythological, biblical, portrait), and technical information (material, size, condition). Even if your instructor has not told you to begin thus, you may find it helpful to start along these lines. The mere act of writing *anything* will probably help you to get going.

Still, almost all writers—professional as well as student writers—find that the beginning paragraphs of their drafts are false starts. In your fin-

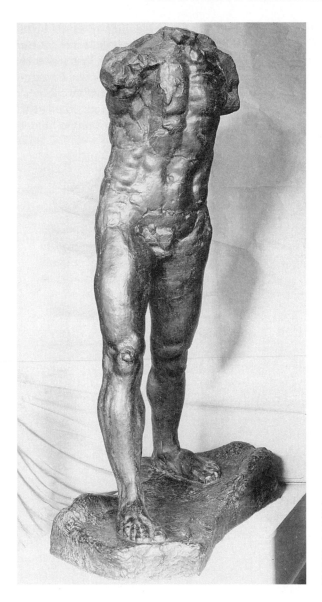

Auguste Rodin,
L'homme qui marche,
1877. Bronze,
7′11¾″. (Rodin
Museum, Paris. Foto
Marburg/Art
Resource, NY)

ished paper the opening cannot be mere throat clearing. It should be interesting and informative. Don't take your title ("Space in Manet's *A Bar at the Folies-Bergère*") and merely paraphrase it in your first sentence: "This essay will study space in Manet's *A Bar at the Folies-Bergère.*" There is no information about the topic here, at least none beyond what the title already gave, and there is no information about you either—that is, there is no sense of your response to the topic, such as might be present in, say,

> The space in A Bar at the Folies-Bergère is puzzling; one at first wonders where the man is standing, whose reflection we see in the upper right.

This brief opening illustrates a surefire way to begin:

- Identify the artwork(s) you will discuss.
- Suggest your thesis in the opening paragraph, moving from a generalization ("The space in the picture is puzzling") to some supporting details ("we are unsure of the position of the man whose reflection we see").

Such an introduction quickly and effectively gets down to business; especially in a short paper, there is no need (or room) for an in-depth introduction.

Notice in the following example, the opening paragraph from an essay on Buddhist art in Japan, how the writer moves from a "baffling diversity" to "one major thread of continuity."

Effective Opening

> Amid the often baffling diversity which appears in so much of the history of Japanese art, one major thread of continuity may be traced throughout the evolution of religious painting and sculpture. This tradition was based on the great international style of East Asian Buddhist imagery, which reached its maturity during the early eighth century in Tang China and remained a strong influence in Japan through the thirteenth century.
>
> —John Rosenfield, *Japanese Arts of the Heian Period: 794–1185*
> (New York: Japan Society, 1967), 23

Similarly, in the example on page 146, the paragraph moves from a general comment about skin-clinging garments to an assertion of the thesis (the convention is exaggerated in English Neoclassical art), and this thesis is then supported with concrete details.

Sample Revised Paragraph

~~The formal elements of a work of art combine to form a dominant impression that contributes significantly to the understanding of the work's meaning.~~ *the principle that in a work of art, form as well as subject matter establishes the meaning.* Two sculptures, ~~located~~ *of seated figures* in the Boston Museum of Fine Arts provide good ~~.~~ examples of ~~this principle.~~ *Dating* from the Fourth Dynasty of Old Kingdom Egypt (2613-2494 B.C.) ~~is~~

a foot-high statue of Prince Khunera, ~~This sculpture,~~ originally placed in his burial site at Giza, depicts the prince sitting cross-legged like a scribe. The yellow limestone sculpture is somewhat chipped and cracked but is generally intact, ~~as is~~ the second work of art, the Bodhisattva of Compassion, also known as Avalokiteshvara. ~~C~~arved from a black stone called schist, ~~the Bodhisattva~~ was made in Bihar, in eastern India, in the mid-tenth century A.D. *The Bodhisattva is more than twice the size of the prince, but* ~~Although more than twice the size of the prince, he sits in a similar~~ *the essential differences are conveyed chiefly by elements of* ~~position. However, both the body language and the overall impres-~~ *design.* ~~sion conveyed by the two sculptures differ greatly. Skillful use of~~ ~~the formal elements of design gives~~ *conveys* Prince Khunera a sense of eternal stillness~~, and~~ *, whereas* the Bodhisattva of Compassion *conveys* a sense of movement and accessibility, ~~making~~ each impression correspond*s* with the religious meaning and purpose of the object.

Student's revision of the opening paragraph of a draft of an essay comparing two sculptures. ☞ The student probably deleted the first sentence because she realized she was lecturing—at length—about a point that her readers would regard as obvious. In the revision she takes what is essential from the first sentence of the draft and incorporates it into the next sentence, thus getting a stronger opening sentence. ☞ In the original version the second sculpture is weakly introduced by being tacked on to the end of a sentence about the first sculpture; in the revision it rightly gets its own sentence. ☞ The sprawling last sentence of the original paragraph is, in the revision, converted into two sentences, allowing the paragraph to end more emphatically.

and sculptural custom of clothing the nude in skin-
many and often-copied Classical precedents, but the
 this convention seems exaggerated in English art of
iod more than at other times and places. The shad-
owy meeting of thighs, the smooth domes of bosom and backside, are
all insisted on more pruriently through the lines of the dress than they
were by contemporary French artists or by Botticelli and Mantegna and
Desiderio da Settignano, who were attempting the same thing in the
Renaissance—or, indeed, than by the Greek sculptors. The popular
artists Rowlandson and Gillray naturally show this impulse most bla-
tantly in erotic cartoons and satirical illustrations, in which women have
enormous bubbly hemispheres fore and aft, outlined by the emphatically sketched lines of their dress.

—Anne Hollander, *Seeing Through Clothes*
(New York: Viking, 1978), 118

Two other effective ways to begin an essay are

- to use a quotation (notice the use of a quotation from Vasari, on
  page 142, at the beginning of the section on introductions)
- to use an interesting relevant fact, such as "Grant Wood's *American Gothic* seems to be so American a painting that it comes as
  a surprise to learn that it is indebted to European sources"

Concluding Paragraphs

The preceding discussion of opening paragraphs quotes Lord Byron:
"Nothing so difficult as a beginning." But he went on to add, "Unless perhaps the end."

In conclusions, as in introductions, try to say something interesting.
It is not of the slightest interest to say "Thus we see . . ." and then go on
to echo the title and the first paragraph. There is some justification for a
summary at the end of a long paper because the reader may have half
forgotten some of the ideas presented thirty pages earlier, but a paper
that can easily be held in the mind needs something different. A good
concluding paragraph does more than provide an echo of what the
writer has already said. It may round out the previous discussion, normally with a few sentences that summarize (without the obviousness of
"We may now summarize"), but usually it also draws an inference that

has not previously been expressed, thus setting the previous material in a fresh perspective. A good concluding paragraph closes the issue while enriching it.

Consider the following example, the concluding paragraph of an essay by a student (Jane Holly) on small gilt bronzes (two or three inches tall) of the Buddha Shakyamuni, produced in Japan in the seventh century AD.

Effective Ending

> Small gilt bronzes of the type that we have been discussing are almost unknown outside of Japan, and they are rare even in Japan. Temples do not have them, since they are too small for public worship, and museums rarely display them, since the casual viewer would pass them by, thinking of them as trifles. When we think of a Japanese Buddha, we probably call to mind some enormous figure, such as the seated Great Buddha at Kamakura--about thirty-four feet tall--which is known throughout the world because airlines often use pictures of it on travel posters. But these small gilded images, doubtless commissioned by wealthy people for use in small shrines in their homes, are by no means trivial. They were obviously made with extreme care, and gilding was a costly, time-consuming process, so the bronzes must have been highly valued. Today, as familiarity with East Asian art increases in America, we are beginning to see the beauty and importance of these physically small but spiritually great works.

Pretty much the same technique is used at the end of Elsen's second paragraph on Van Dyck's *Charles I* (page 141), where the writer moves from a detailed discussion of the picture to the assertion that the picture continues to attract us.

Don't feel that you must always offer a conclusion in your last paragraph. When you have finished your analysis, it may be enough to stop— especially if your paper is fairly short, let's say fewer than five pages. If, for example, you have throughout your paper argued that a certain Japanese print shows a Western influence in its treatment of space, you need scarcely reaffirm the overall thesis in your last paragraph. Probably it will be conclusion enough if you just offer your final evidence, in a well-written sentence, and then stop.

Speaking of well-written sentences, an effective quotation, perhaps by the artist you are writing about, may provide a strong ending to your essay.

✔ **Checklist for Revising Paragraphs**

✔ Does the paragraph *say* anything? Does it have substance?

✔ Does the paragraph have a topic sentence? If so, is it in the best place? If the paragraph doesn't have a topic sentence, might one improve the paragraph? Or does it have a clear topic idea?

✔ If the paragraph is an opening paragraph, is it interesting enough to attract and to hold a reader's attention? If it is a later paragraph, does it easily evolve out of the previous paragraph and lead into the next paragraph?

✔ Does the paragraph contain some principle of development, for instance from cause to effect, or from general to particular?

✔ Does each sentence clearly follow from the preceding sentence? Have you provided transitional words or cues to guide your reader? Would it be useful to repeat certain key words for clarity?

✔ What is the purpose of the paragraph? Do you want to summarize, to give an illustration, to concede a point, or what? Will your purpose be clear to your reader, and does the paragraph fulfill your purpose?

✔ Is the closing paragraph effective and not an unnecessary restatement of the obvious?

A NOTE ON TENSES

In talking about an artist's life, you will probably use forms of the past tense ("Georgia O'Keeffe *was born* in Sun Prairie, Wisconsin. She *taught* in Texas"). But in speaking about works of art themselves it is usually best to use the present tense: "Her work *hangs* in many museums," "This painting *shows* her at her best," "The blue *is* intense," "She *denied* that she used sexual symbolism, but her pictures nonetheless *contain* sexual symbols." Some instances can go either way. Compare two sentences:

> Her usual motifs *are* Southwestern, such as desert flowers and bleached bones.

> Her usual motifs *were* Southwestern, such as desert flowers and bleached bones.

The first sentence uses the present tense because the writer is thinking of the works as enduring presences—they are here right now, and the motifs are Southwestern. The second sentence uses the past tense because the writer is telling us about what O'Keeffe *did*—she often chose Southwestern motifs. What a writer cannot do, however, is to shift back and forth. If you write, "Her usual motifs *were* Southwestern," you almost surely cannot write in your next sentence, "She often *paints*. . . ."

Of course you can use the *narrative present* (also called the *historic present*): "Gothic carvers delight in ornament," "Van Gogh usually applies the pigment boldly." (This device is commonly said to add vitality to the narrative.) But if you use the narrative present—and it often sounds odd—be consistent. If you write, "Gothic sculptors *delight* in ornament; they *carved* abundant tendrils, and they *fill* every corner," you will needlessly disturb your readers, who will wonder why you use a past tense ("carved") between two verbs in the present tense ("delight" and "fill").

6

Some Critical Approaches

Most of this book thus far has been devoted to writing about what we perceive when we look closely at a work of art, but it is worth noting that other kinds of writing can also help a reader to see (and therefore to understand better) a work of art. For example, one might discuss not a single picture but, say, a motif: Why does the laborer become a prominent subject in nineteenth-century European painting? Or: How does the theme of leisure differ between eighteenth-century and nineteenth-century painting; that is, what classes are depicted, how aware of each other are the figures in a painting, what are the differences in the settings and activities—and why? Or: Why do Europeans and Americans on the whole put a much higher value on nineteenth-century African art than they do on African art of the second half of the twentieth century?

This chapter sketches some of the chief methodologies or ways of approaching art—for instance, by setting it in its historical context or by examining the psychology of the artist—but it is important to understand that these approaches are not mutually exclusive. It's not a question of one method *or* another; art historians use all of the tools they feel comfortable with. There are times when, so to speak, one stands back from the work, and times when one gets up close, times when one uses a telescope and times when one uses a microscope. What follows is an introduction to some of the methodologies or sets of principles that art historians have found helpful when they study art.°

°For a collection of writings about art, from Vasari (sixteenth century) to Griselda Pollock (late twentieth century), see *Art History and Its Methods,* ed. Eric Fernie (London: Phaidon, 1995). Fernie also includes a readable glossary (about forty pages) of terms ranging from *art* and *avant garde* through *typology* and *vocabulary,* with mini-essays on each term. For a glossary with more terms but with briefer definitions, see Paul Duro and Michael Greenhalgh, *Essential Art History* (London: Bloomsbury, 1992).

SOCIAL HISTORY: THE NEW ART HISTORY AND MARXISM

Discussions of subject matter may be largely **social history,** wherein the aesthetic qualities of the work of art (as well as such matters as whether a work is by Rembrandt or by a follower) may be of minor importance. Social historians assume that every work (if carefully scrutinized) tells a story of the culture that produced it. Thus, Gary Schwartz, in *Rembrandt: His Life, His Paintings* (1985), says that his intention is to study Rembrandt as "an artistic interpreter of the literary, cultural, and religious ideas of a fairly fixed group of patrons" (page 9).

Notice the interest in patrons. Much social history is interested in patronage, confronting such questions as "Why did portraiture in Italy in the late fifteenth century show an increased interest in capturing individual likeness?" and "Why did easel painting (portable pictures) come into being when it did?" The social historian usually offers answers in terms of who was paying for the pictures. Obviously a Renaissance duke or the wealthy wife of a nineteenth-century businessman wanted images very different from those wanted by the medieval church.

Let's consider what the social historian's point of view may reveal if we scrutinize, say, French "Orientalist" paintings—nineteenth-century pictures of North Africa (e.g., Algeria and Morocco) and the Near East by such artists as Delacroix and Gérôme, painted for the middle-class European market. We might find that these paintings of part of the Islamic world, such as Gérôme's *The Slave Market* (see page 152), do not simply depict locales; rather, the paintings also depict the colonialist's Eurocentric view that this region is a place of savagery and abundant sexuality (beautiful slaves, male and female). The pictures offered viewers (or at least some of them) voyeuristic pleasure by providing images of carnal creatures, and at the same time the pictures allowed the viewers comfortably to feel that the region was badly in need of European law and order. In short, they can be seen as European constructions that justify colonialism, for instance the French occupation of Algeria (1830) and the British occupation of Egypt (1882). Even Matisse's pictures of barebreasted odalisques reclining on cushions and staring blankly can reasonably be said, by today's standards, to embody the colonialist's views, or at least the early twentieth-century bourgeois male's views of North Africa and of women as sensuous and passive beings waiting to be enlivened by the white male who gazes at the pictures. And of course white males

Jean-Léon Gérôme, *The Slave Market*, c. 1867. Oil on canvas, 33³⁄₁₆″ × 24¹³⁄₁₆″. Gerald M. Ackerman, writing in *Arts Magazine* 60 (March 1986): 79, says that this picture of a slave market "could only be seen as abolitionist." Richard Leppert, in *Art and the Committed Eye* (Boulder: Westview, 1996), 239, emphatically disagrees: "We are *not* invited by Gérôme to gaze at this image in order to critique slavery; we are invited to look and to salivate, all the while saying out loud to anyone within earshot, 'Tsk, tsk.'" (Sterling and Francine Clark Institute, Williamstown, Mass.)

were the chief purchasers. (We will return to this idea—that *the gaze* of the viewer implies power over the depicted subject—when we discuss images of female nudes as viewed by males later in the chapter.)*

In this view, then, a work of art (like a religious, legal, or political system) is a creation deeply implicated in the values of the culture that produced and consumed it. The work is a "text" whose attractive surface is a mask. The student, by resisting its aesthetic seductions and perhaps by "reading against the grain" (a common expression referring to the attempt to get at what the artist may have tried to hide or may have been unaware of), "interrogates" or "demystifies" or "deconstructs" the work. The approach, which denies that the individual artist establishes the meaning of a work, is called **deconstruction.**

The study of the social and political history of art, especially of matters of class, gender, and ethnicity, and especially when conducted in a somewhat confrontational manner, is sometimes called the **New Art History** to distinguish it from the earlier art history that was concerned chiefly with such matters as biography, stylistic development, attribution, aesthetic quality, and symbolic meanings of subjects. This new approach was summed up more than twenty-five years ago in Kurt Forster's "Critical History of Art, or Transfiguration of Values?" published in *New Literary History* (1972). Dismissing the traditional art history, Forster argued that because its practitioners admired the objects they studied, these art historians could not study them critically—that is, they could not see that the objects served the vested interests of the classes in power. Calling for a new approach, Forster argued that "the only means of gaining an adequate grasp of old artifacts lies in the dual critique of the ideology which sustained their production and use, and of the current cultural interests that have turned works of art into a highly privileged class of consumer and didactic goods" (pages 463–64).

In this view, we must examine the economic and cultural demands made in the world in which the artist worked, and we must examine ourselves—taking account of the influence of such factors as our race, class, and sex—to understand why we regard certain works as of special value. This kind of analysis, which studies a work (often called a "text") in terms

---

*For a survey of views concerning Orientalism, see John M. MacKenzie, *Orientalism: History, Theory and the Arts* (Manchester: Manchester University Press, 1995). For a severe view of the Orientalists, see Linda Nochlin's essay, "The Imaginary Orient," *Art in America* 71 (1983): 186–91, reprinted in her book *The Politics of Vision* (New York: Harper, 1989), 33–59. See also *The Orientalists: Delacroix to Matisse: The Allure of North Africa and the Near East*, ed. Mary Anne Stevens (London: Royal Academy of the Arts, 1984), and (on Victorian photographs) James R. Ryan, *Picturing Empire* (1998).

of the conditions of its production and reception, is commonly called **cultural materialism** or **cultural criticism.** The interest is less in aesthetic judgment than in moral or political judgment, especially in matters of race, gender, and class. Thus, cultural materialists, influenced by Marxism, are far less likely to ask, "Is this work by Rembrandt or by a follower" (a question that probably gets into matters of quality) than they are to ask, "What is the social system that causes people to value productions of this sort?" and, "How does this object sustain—or undermine—the prevailing power relations?" Similarly, confronted with a Renaissance painting of an angel facing a young woman above whose head is a halo, cultural materialists are likely to be more interested in an Oriental rug on a table ("In sixteenth-century Europe, rugs of this sort were expensive and conveyed high status, therefore they were much coveted by the aristocracy and the wealthier members of the bourgeoisie") than in the ostensible subject matter, the Annunciation (the announcement by the angel Gabriel to the Virgin Mary, "You shall conceive and bear a son, and you shall give him the name Jesus").

A chief goal of the New Art History, then, is to unearth or reconstruct the often neglected, forgotten, or unperceived ideological assumptions or social meanings that inform artworks. A related goal is to show that works of art not only reflect ideology—and therefore are far more than mere examples of beauty—but also actively participate in ideological conflicts. In this view, every work is political; in effect it says to the viewer, "See things *this* way, not some other way." Whether or not the men and women who fashioned the works are politically aware, the works produce political actions.°

°For a collection of essays along these lines, see A. L. Rees and Frances Borzello, eds., *The New Art History* (London: Camden, 1986). For more advanced discussions, see Norman Bryson, Michael Ann Holly, and Keith Moxey, eds., *Visual Culture: Images and Interpretations* (Hanover: Wesleyan UP 1994), and an issue of a journal, *Critical Inquiry* 15 (Winter 1989). Inevitably, new critical approaches bring new critical terms. A few terms, such as *cultural materialism, deconstruction, Orientalism,* and *text* have been used in the present chapter, but dozens of others are now in use, some (such as discourse, text, and queer theory) whose meanings are not as simple as they seem to be at first glance. The only way to understand these terms is to read a good deal of contemporary criticism, thereby absorbing the terms within their contexts, but considerable help can be found in the glossaries in the books mentioned in the footnote on page 150, and also by consulting Thomas Patin and Jennifer McLerran, *A Glossary of Contemporary Art Theory* (Westport, Conn.: Greenwood, 1997). The *Glossary* offers definitions ranging from two or three sentences to a page or two. For essays (not always easily comprehended) on such newly popular terms as *gaze, postcolonialism,* and *simulacrum,* see *Critical Terms for Art History,* ed. Robert S. Nelson and Richard Schiff (Chicago: U of Chicago P, 1996).

The interest in patrons is only one aspect of the social historian's concern with the production and consumption of art. Other points of interest include the particular artists in any given period and the sorts of training they received. In *Art News* (January 1971), Linda Nochlin asks a fascinating sociological question: "Why Have There Been No Great Women Artists?" (The essay is reprinted in Nochlin's book, *Women, Art, and Power and Other Essays,* 1988.) Nochlin rejects the idea that women lack artistic genius, and instead she finds her answer "in the nature of given social institutions." For instance, women did not have access to nude models (an important part of the training in the academies), and women, though tolerated as amateur painters, were rarely given official commissions. And, of course, women were expected to abandon their careers for love, marriage, and the family. Furthermore, during certain periods, women artists were generally confined to depicting a few subjects. In the age of Napoleon, for example, they usually painted scenes not of heroism but of humble, often sentimental domestic subjects such as a girl mourning the death of a pigeon.

The social historian assumes that works of art carry ideas and that these ideas are shaped by specific historical, political, and social circumstances. The works are usually said to constitute ideologies of power, race, and gender. Architecture especially, being made for obvious uses (think of castles, cathedrals, banks, museums, schools, libraries, homes, malls), often is usefully discussed in terms of the society that produced it. (Architecture has been called "politics in three dimensions.")

Social historians whose focus is **Marxism** examine works of art as reflections of the values of the economically dominant class and as participants in political struggles. Works of art themselves do some sort of work. For Marxists this work usually is, at bottom, the reinforcement of the class ideology. In this view, the meanings and value of a work can be understood only by putting it into the social situation that produced it.* In the words of Nicos Hadjinicolaou, in *Art History and Class Struggle* (1978), "Pictures are often the product in which the ruling classes mirror themselves" (page 102). Of course the mirror is not really a mirror but a presentation of the ways in which the ruling classes wish to be seen. For instance, the landowners whose wealth is derived from the soil, or more

---

*See Janet Wolff, *The Social Production of Art* (New York: St. Martin's, 1981). Among important Marxist writings on art are John Berger, *Ways of Seeing* (London: Penguin, 1972), T. J. Clark, *The Painting of Modern Life* (New York: Knopf, 1985), and Terry Eagleton, *The Ideology of the Aesthetic* (Oxford: Blackwell, 1990). For a comment about Marxism and the evaluation of art, see page 178.

precisely from the agricultural labor of peasants, may present themselves as in harmony with nature, or as the judicious and loving caretakers of God's earth. Artworks thus are masks; their surface is a disguise, and the Marxist art historian's job is to discover the underlying political and economic significance. Other artworks, instead of reinforcing a society's dominant values, work the other way, that is, they subvert the values, but in either case art is part of an ideological conflict.

Where does beauty come in? To the question, "Why does this work give me pleasure?" the Marxist historian—or at least Hadjinicolaou—replies in this way:

> Aesthetic effect is none other than the pleasure felt by the observer when he recognizes himself in a picture's visual ideology. It is incumbent on the art historian to tackle the tasks arising out of the existence of this recognition. . . . This means that from now on the idealist question "What is beauty?" or "Why is this work beautiful?" must be replaced by the materialist question, "By whom, when and for what reasons was this work thought beautiful?"
>
> —Nicos Hadjinicolaou, *Art History and Class Struggle* (London: Pluto, 1978), 182–83

GENDER STUDIES: FEMINIST CRITICISM AND GAY AND LESBIAN STUDIES

Gender studies, a comprehensive field including feminist and gay and lesbian historical scholarship and criticism, attempts to link all aspects of cultural analysis concerning sexuality and gender. These methodologies usually assume that although our species has a biologically fixed sex division between male and female (a matter of chromosomes, hormones, and anatomical differences), the masculine and feminine roles we live out are not "natural" or "essential" or "innate" but are established or "constructed" by the society in which we live. Society—or cultural interpretation—it is argued, exaggerates the biological sexual difference (male, female) into gender difference (masculine, feminine), producing ideals and patterns of gender (masculinity, femininity) and of sexual behavior (e.g., the idea that heterosexuality is the only natural behavior). In our patriarchal culture, it is argued, parents, siblings, advertisements, and so forth teach us that males are masculine (strong, active, rational) and that females are feminine (weak, passive, irrational), and we (or most of us) play male or female roles in a social performance, "constructing" ourselves into what society expects of us. Simone de Beauvoir, an early exponent of the feminist theory of gender, in *The Second Sex* (1949) puts it this way: a "Woman" is constructed as "Man's Other," thus "one is not born a woman; one becomes one." Because (according to the view that gender is socially constructed) ideals and patterns change from one time or culture to another, analysis based on this awareness can help the viewer better understand works of art by or about men and women, heterosexual and homosexual.°

Feminist art history and criticism begins with the fact that men and women are different. As we have just seen, some writers believe the difference is chiefly "essential"—for instance, women menstruate and bear children—whereas others believe the difference is chiefly "constructed." In Mary D. Garrard's view, for example,

> the definitive assignment of sex roles in history has created fundamental differences between the sexes in their perception, experience and expectations of the world, differences that cannot help but be carried over into the creative process where they have sometimes left their tracks.
> —*Artemisia Gentileschi: The Image of the Female Hero in Italian Baroque Art* (Princeton: Princeton University Press, 1989), 202

One can make further distinctions, arguing, for instance, that the experiences of women of color differ from those of white women, but here we will concern ourselves only with the most inclusive kind of feminist writing. This writing is especially interested in two topics: (1) how women are portrayed in art and (2) whether women (because of their biology or socialization or both) create art and see art differently from men. The first topic is centered on subject matter: How do images of women (created chiefly by men) define "being female"? Are women depicted as individuals with their own identity, or chiefly as objects for men to consume?

The second topic, women as artists and viewers of art, is closely related to the first. It is probably true that in a world of predominantly male-created public works of art, the implied viewer is male. Consider pictures of nude females. To use language introduced on page 35, the

---

°On the essentialist/constructionist issue, see Amelia Jones's introductory essay in an exhibition catalog, *Sexual Politics* (Los Angeles: U of California P, 1996), edited by Jones.

bearer of the *gaze* (and therefore the one who wields the power) is male; the eroticized object of the gaze, the "powerless Other," is female. (Erotic gazing is sometimes called *scopophilia,* "pleasures of the eye," a word borrowed from Freud.) The nude woman may be unaware that she is being spied on—perhaps she is about to bathe—or she may be aware, in which case she may be shielding her breasts or her genitals, an action that only emphasizes her vulnerability. In any case, the female nude is an object created largely for the enjoyment of men. Take Picasso's *Les Demoiselles d'Avignon* (see page 26), a brothel scene. One critic has said:

> Everything about the *Demoiselles,* and most of all the brothel situation it forces on the viewer, designates that viewer as male. To pretend . . . that women can play the customer—in *that* brothel in particular—or that as gendered subjects they can relate to the work on terms equal to men denies the intensity, force, and vehemence with which Picasso differentiated and privileged the male as viewer.

In this assertion true? All viewers will, obviously, have to form their own answers.

Why, one can ask, do some heterosexual women take pleasure in some images of female nudity? Here are three of the many answers that have been offered: (1) these women, socialized by a patriarchal culture

Guerrilla Girls, *Do Women Have to Be Naked?,* 1989. Poster, ink and color on paper, 11″ × 28″. (Private collection. Courtesy of Guerrilla Girls) Since 1985 a group of women, wearing gorilla masks, have put up posters and issued other materials calling attention to the underrepresentation of women in the art world. The image on this poster is based on Ingres's *Grande Odalisque* (1814), a painting that few viewers would deny offers an eroticized woman as the object of the gaze. Interestingly, the painting was commissioned by a woman, Queen Caroline Murat.

(i.e., deceived or coerced into accepting male ideologies), may negate their own experience and identify with the heterosexual, masculine, voyeuristic, penetrating gaze; (2) some female viewers may narcissistically identify themselves with the female images depicted (this explanation is sometimes used to account for images of female nudity found in advertising directed toward women); (3) "being female is only one aspect of a woman's experience," and although it may sometimes determine her response as a viewer, at other times it may not. (This third view is offered by Carol Ockman, in *Ingres's Eroticized Bodies*, 1995.)

How female viewers respond to Degas's images of nude women bathing themselves, often in highly contorted positions, has recently evoked abundant commentary. For Griselda Pollock, in *Dealing with Degas* (1992), these images show "obsessive, repetitious re-enactments of sadistic voyeurism. . . . The model's body [is] debased and abused," and a female viewer is "forced to take up the proferred sadistic masculine position and symbolically enact the violence of Degas's representation, or identify masochistically with the bizarrely posed and cruelly drawn bodies" (page 33). But, to take only one countervoice, Wendy Lesser in an essay called "Degas's Nudes" (in *His Other Half: Men Looking at Women Through Art*, 1991), sees empathy rather than aggression, and sensuousness rather than sadism. Lesser argues that the woman viewer is encouraged to identify herself with these deeply absorbed unselfconscious women bathing themselves.*

To turn to women as artists: Are the images created by women different from those created by men? (The term *gynocriticism* is sometimes used to refer to the branch of feminist criticism that is especially concerned with recovering from obscurity the work of women artists and carefully examining it.) In working in the "masculine" genre of the female nude, does a female artist (e.g., Suzanne Valadon) produce images that significantly differ from the images produced by males working in the same artistic milieu? (The images produced by males are usually said to be passive objects, constructed by male desire.) Are there certain genres in which women chiefly worked? Are certain traits that are said to characterize the work of many women artists—e.g., sentimentality—not weaknesses but strengths, evidence of a distinct way of resisting the male status quo?

---

*For more about the gaze, see Marcia Pointon, *Naked Authority: The Body in Western Painting, 1830–1908* (Cambridge: Cambridge University P, 1990).

Some writers have emphasized connections between events in the lives of women artists and their work. For instance, we know that Artemisia Gentileschi was raped by one of her father's apprentices. Some scholars have suggested that Gentileschi's depiction of Judith, the beautiful widow who according to the Bible saved the Hebrews by beheading Holofernes, an Assyrian general, was Gentileschi's way of taking revenge on men. Other writers, however, have pointed out that the subject was painted by men as well as by women, and still others have expressed uneasiness over the tendency to interpret the work of women in terms of their lives. Mary D. Garrard, in *The New York Times* (22 September 1991), calls attention to the implications in some biographical studies: "The art of women is defined in private and personal terms while the art of men is elevated to the level of 'universal' expression."

A glance at certain writings about Frida Kahlo (1907–54), the politically radical painter, supports Garrard's comment. Kahlo suffered greatly, both physically (at the age of eighteen she was partly paralyzed by a horrendous traffic accident in which her spine was fractured, her pelvis was crushed, and one foot was broken) and mentally (her husband, the painter Diego Rivera, was notoriously unfaithful). Discussions of Kahlo's work usually emphasize the connection of the imagery with her suffering, and they tend to neglect the strong political (Marxist and nationalistic) content.

Speaking of "strong" ideas, keep in mind that such words as *strong, energetic, vigorous, forceful, seminal, potent,* and *powerful,* when used to describe artworks, are not gender-neutral but are loaded in favor of male values. Indeed, much of the language of art criticism—*masterpiece* is another example—tends to put women and their work at a disadvantage. *The Guerrilla Girls' Bedside Companion to the History of Western Art,* a book that appropriately describes itself as "a wildly entertaining and much-needed corrective to traditional art history," quotes (page 41) a telling example by the art historian James Laver:

> Some women artists tend to emulate Frans Hals, but the vigorous brush strokes of the master were beyond their capability. One has only to look at the work of a painter like Judith Leyster to detect the weakness of the feminine hand.

Today we rub our eyes in amazement that such stuff passed as serious art history. It is easy to see why many art historians now argue that the traditional distinction between "high art" (a portrait of Henry VIII by Holbein) and "low art" (a quilt by an anonymous woman) masquerades as a universal truth, but it is really a patriarchal concept designed to devalue female creativity.°

Gay and lesbian art criticism, like feminist criticism, operates from the principle that varieties of sexual orientation make important differences in how artists portray the world, love, and Eros, and in how viewers receive and interpret those images.†

°Examples of feminist criticism and historical scholarship can now be found in almost all journals devoted to art, but they are especially evident in *Women and Art, Woman's Art Journal,* and *Women's Art.* For collections of feminist essays, see Norma Broude and Mary D. Garrard, eds., *Feminism and Art History* (New York: Harper, 1982); Arlene Raven, Cassandra L. Langer, and Joanna Frueh, eds., *Feminist Art Criticism* (Ann Arbor: UMI Research, 1988); and Norma Broude and Mary D. Garrard, eds., *The Expanding Discourse: Feminism and Art History* (New York: Harper, 1992). For a survey of feminist art history, see Thalia Gouma-Peterson and Patricia Mathews, "The Feminist Critique of Art History," *Art Bulletin* 69 (1987): 326–57; and the follow-up by Norma Broude, Mary D. Garrard, Thalia Gouma-Peterson, and Patricia Mathews, "An Exchange on the Feminist Critique of Art History," *Art Bulletin* 71 (1989): 124–27. For a valuable survey of feminist methods and themes, see Lisa Tickner, "Feminism, Art History, and Sexual Difference," *Genders* 3 (1988): 92–129. For a listing of feminist writing about art— including more than a thousand items, with brief summaries—see Cassandra Langer, *Feminist Art Criticism: An Annotated Bibliography* (New York: G. K. Hall, 1993).

†This discussion of gay and lesbian criticism is by James M. Saslow (Queens College, City University of New York), author of numerous studies including *Ganymede in the Renaissance: Homosexuality in Art and Society* (New Haven, Conn.: Yale University Press, 1986). Some critics use the term **queer studies** as an umbrella term that covers not only lesbian and gay studies, each of which has its own history, but also bisexual and transgender studies. Some writers think of gay studies as largely biographical, and queer studies (or **queer theory**) as largely theoretical—the theory being (usually) the view that the self has no fixed identity but is a social construction: Sex (usually a matter of two categories, male and female) is biological, but masculinity and femininity are (as was said on pages 156–57) fluid social constructions, mere fictional images. In this view, because we all see things from a unique position, we are all rather queer—psychologically incomprehensible to others— though society seeks to regulate gender and to enforce a system of compulsory heterosexuality. Indeed, this view that heterosexual relationships are as varied, fluid, and socially transgressive as same-sex relationships is an important aspect of queer culture—a term much used but without much agreement about its meaning. Finally, many writings that announce themselves as examples of queer theory are not in fact theoretical, since they do not offer a general principle as a way of explaining phenomena. These writings, perhaps more accurately characterized as queer studies, are especially concerned with investigating how desire is formed, expressed, and perceived. See Whitney Davis, "'Homosexualism,' Gay and Lesbian Studies, and Queer Theory in Art History," in *The Subjects of Art History,* eds. Mark A. Cheetham, Michael Ann Holly, and Keith Moxey (Cambridge: Cambridge University Press, 1998); Annamarie Jagose, *Queer Theory: An Introduction* (Carlton, Vic.: Melbourne UP, 1996); *Feminism Meets Queer Theory,* ed. Elizabeth Weed and Naomi Schor (Bloomington: Indiana University Press, 1997).

Again like feminist art criticism and historical scholarship, gay and lesbian criticism faces certain problems of definition and scope. Most broadly, it means both art *about* gay men and lesbians (or, more generally, homosexuality and bisexuality in all their various forms) and art *by* gay and lesbian artists. "Gay art" is not synonymous with erotic art or pornography. It can include, for example, genre scenes of gay life (Francis Bacon, David Hockney), portraits of famous individuals who were gay or lesbian (Berenice Abbott's photographs of the literary lesbians of Paris in the period between the two world wars), political art around social issues of special concern to lesbians and gay men (today, AIDS posters), and mythological or historical subjects (Jupiter and his cupbearer Ganymede, the lesbian poet Sappho).

Gay and lesbian art writing looks at homosexuality both as subject matter/text and as a factor that may help to shape the work of the individual artist or creator. It asks about differences or parallels in the ways that love, sexuality, gender, and daily life are depicted by heterosexuals and homosexuals; and it seeks to uncover distinctive expressions of the gay and lesbian experience that have, in the past, often been suppressed or misunderstood (e.g., Greek vase paintings of homosexual courtship, Caravaggio's mythologized portraits of Roman boys, Rosa Bonheur's self-representation in male dress). It should be emphasized, however, that not all art by gay people (or by heterosexual men or women) can be reduced to biographical illustration.

While biography and psychology are important tools in understanding some art by gays and lesbians, gay art criticism is linked equally closely to both political and social history. Political because, throughout much of Western history at least, homosexual expression, when not silenced, has often been forced to operate indirectly, in a sort of code. Moreover, while heterosexuals have also represented scenes of homosexual life (for example, Brueghel's *Village Kermesse at Hoboken,* with its male couple embracing at a carnival), their "outsider" images have often propagated negative attitudes that need to be placed in historical perspective.

In confronting images, it is important to avoid projecting one's own moral attitudes back onto earlier art. One should not assume, for example, that all cultures "naturally" condemn homosexuality; rather, one should try to discern what artists were trying to say within their own cultural framework. Gay art criticism is thus closely linked to social history, in that understanding of art about gay experience requires an understanding of attitudes about sex and gender in various societies. Indeed, not only attitudes but also the very definitions of sexual identities and roles have varied greatly throughout history. "Homosexuality" is itself a modern Western term, and vase-paintings of male adolescents addressed to Greek men (who were generally what we would now call bisexual), or woodblocks showing two women making love in China (where no equivalent of a contemporary lesbian culture existed), need to be understood in terms of the categories of sexual experience and the social patterns constructed by those cultures.

Much of what has just been said concerns iconography (see page 166); in formal and visual terms, the issue of *the gaze*, adopted from feminist criticism, is important in gay and lesbian analysis as well. By "the gaze" I mean the act of looking itself, including both how it is established within works of art and what kind of outside viewer is implied. Students should ask such questions of pictures as "*Who* is doing the looking?" and "*Who* or *what* is being looked at?" Consider, for example, the different possible meanings of a male figure when he is painted or looked at by different viewers:

- a heterosexual male artist, for whom the sitter and his body are primarily an opportunity for scientific study, a traditional subject for drawing class, and the like
- a gay male artist, for whom the subject also has an erotic interest and perhaps some significant biographical connection to the painter (e.g., Michelangelo's drawings of handsome young men and mythological boys)
- a heterosexual woman, who, like a gay man, may also be erotically interested in the figure (the question of why there are no "academic" drawings by women before the twentieth century is a related question for feminist history as well)
- a lesbian artist, who, like the heterosexual male, has no erotic interest in the sitter, but who might feel political or cultural solidarity (e.g., Romaine Brooks's portrait of Jean Cocteau)*

BIOGRAPHICAL STUDIES

Biographical writing, glanced at in the preceding comments on gender criticism, need not be rooted in gender. True, one might study the degree to which Picasso's style changed as he changed wives or mistresses,

---

*For a guide to the existing literature in this rapidly expanding field, consult *Bibliography of Gay and Lesbian Art* (New York: College Art Association, Gay and Lesbian Caucus, 1994). A second edition, edited by Ray Anne Lockard, is in preparation.

but one could also study his changes in style as he changed literary friends—Apollinaire affecting Picasso's Cubism, Cocteau his Neoclassicism, and Breton his Surrealism. In fact, John Richardson's *A Life of Picasso* is built on a remark by Picasso: "My work is like a diary. To understand it, you have to see how it mirrors my life."

Speaking broadly, we can say that biographical studies are of two sorts, those that emphasize the individual artist's genius ("the life of") and those that emphasize the artist's social context ("the life and times of"). The first kind is as old as Vasari, whose *Lives of the Artists* (1550) is traditionally said to be the first modern example of art history. Vasari emphasizes the mysterious genius of the individual (in Renaissance Italy, all were white and almost all were males): Giotto was a child prodigy who later was able to draw a perfect circle freehand, Michelangelo was a God-like creator, etc. The second kind of biographical study, emphasizing the social context, sees the artist not as a mysterious genius but as someone efficiently producing a commodity in response to the market. In response to the charge that biographies depend at least in part on the chance survival of written documents, defenders of life-and-times biographical studies argue that they set the works of art in their original historical and authorial contexts, and are concerned with the issues that most concerned the artists themselves—such things as "Who is paying?" and "How is the workshop organized?" Still, readers may feel that the traditional biographical-historical approach to art history conveys very little sense of why we today value these earlier works of art.

PSYCHOANALYTIC STUDIES

Many of today's biographical studies can be called **psychoanalytic studies.** Such writings follow in the tradition of Sigmund Freud's *Leonardo da Vinci: A Psychosexual Study of Infantile Reminiscence* (1910), which seeks to reconstruct Leonardo's biography and psychic development by drawing on certain documents and, especially, by analyzing one of Leonardo's memories (for Freud, a fantasy) of an experience while he was an infant. Thus, it seems that Leonardo had two "mothers," a biological mother (a peasant woman) and a stepmother (the woman who was Leonardo's father's wife). In accord with this information, Freud sees in *The Madonna, Child, and St. Anne* Leonardo's representation of himself as the Christ Child, the peasant woman as St. Anne, and the stepmother as the Madonna. St. Anne's expression, in Freud's analysis, is both envious (of the stepmother) and joyful (because she is with the child whom she bore).

In a 1931 letter Freud wrote that an artist's entire repertory might be traced back to "the experience and conflicts of early years," and most psychoanalytic studies of artists have followed Freud in concentrating on early experiences. But there are exceptions; one example is Henry Adams's psychoanalytic essay (in *Art in America*, February 1983) on Winslow Homer, which argues that the works are related to crises in Homer's life. For instance, *The Life Line* and *The Undertow*, two pictures that show men rescuing women, are said to reveal Homer's love for his mother and his desire to rescue her from death. While we are talking about Homer, Nicolai Cikovsky, Jr., in *Winslow Homer* (1996) says that "The sharks in *The Gulf Stream*, for example, circling the helpless boat with sinuous seductiveness, can be read as castrating temptresses, their mouths particularly resembling the *vagina dentata*, the toothed sexual organ that so forcefully expressed the male fear of female aggression" (page 379). In reading psychoanalytic interpretations, it is well to remember a comment attributed to Freud: "Sometimes a cigar is just a cigar."

Psychoanalytic interpretations of works of art, then, are likely to see the artwork as a disguised representation of the artist's mental life, and especially as a manifestation of early conflicts and repressed sexual desires. The assumption is that the significant "meaning" of the work is the personal meaning that it had (consciously or unconsciously) for the artist. However, as we have seen (pages 17–18), this view is not popular today.

Further, in some psychoanalytic studies the works of art almost disappear. Many pages have been written about why van Gogh cut off the lower part of his left ear and took it to a brothel, where he gave it to a prostitute with the request that she "keep this object carefully." Among the theories offered to explain van Gogh's actions are the following: (1) Van Gogh was frustrated by the engagement of his brother, Theo, and by his failure to establish a working relationship with Paul Gauguin. He first directed his aggression toward Gauguin, and then toward himself. (2) Van Gogh identified himself with prostitutes as social outcasts. He had written that "the whore is like meat in a butcher shop," and so now he treated his body like a piece of meat. (3) He experienced auditory hallucinations, and so he cut off a part of his ear, thinking it was diseased.°

°On van Gogh, see William M. Runyan, *Life Histories and Psychobiography* (New York: Oxford University Press, 1982), 38–41. For a survey of psychoanalytic scholarship and art, see Jack Spector, "The State of Psychoanalytic Research in Art History," *Art Bulletin* 70 (1988): 49–76. On Leonardo, see Bradley L. Collins, *Leonardo, Psychoanalysis, and Art History* (Evanston: Northwestern University Press, 1997).

Although most psychoanalytic studies of art have been concerned with the process of artistic creation (the study of Winslow Homer is an example), in recent years psychoanalysis has been interested in a second area, the source of the perceiver's aesthetic enjoyment. Studies of this topic, often couched in terms of *the gaze*, are primarily concerned with how differences in pleasure are linked to differences in gender and class. (See pages 151–52, 158–59, 163.)

ICONOGRAPHY AND ICONOLOGY

One kind of highly specialized study that keeps the image in view is **iconography** (Greek for "image writing"), or the identification of images with specific symbolic content or meaning. In Erwin Panofsky's words, iconography is concerned with "conventional subject matter," the iconographer showing us that certain forms are (for example) particular saints or gods or allegories. An iconographic study might point out that a painting by Rembrandt of a man holding a knife is not properly titled either *The Butcher* or *The Assassin*, as it used to be called, since it depicts St. Bartholomew, who, because he was skinned alive with a knife, is regularly depicted with a knife in his hand. (Saints, like pagan gods, often hold an *attribute*—that is, some object that serves to identify them. St. Peter, e.g., holds keys, in accordance with Jesus's words to Peter, as reported in Matthew 16:19: "I will give you the keys of the kingdom.")

But is every image of a man holding a knife St. Bartholomew? Of course not, and if, as in Rembrandt's painting, the figure wears contemporary clothing and has no halo, skepticism is warranted. Why can't the picture be of an assassin—or of a butcher or a cook or a surgeon? But a specialist in the thought (including the art) of the period might point out that the identification of this figure with St. Bartholomew is reasonable on two grounds. First, the picture, dated 1661, seems to belong to a series of pictures of similar size, showing half-length figures, all painted in 1661, and at least some of these surely depict apostles. For example, the man with another man at his right is Matthew, for if we look closely, we can see that the trace of a wing protrudes from behind the shoulder of the accompanying figure, and a winged man (or angel) is the attribute or symbol of Matthew. In another picture in this series a man leaning on a saw can reasonably be identified as Simon the Zealot, who was martyred by being sawn in half. Second, because the Protestant Dutch were keenly interested in the human ministry of Christ and the apostles, images of the apostles were popular in seventeenth-century Dutch art, and it is not surprising that Rembrandt's apostles look more like solid citizens of his day than like exotic biblical figures. In short, to make the proper identification of an image, one must understand how the image relates to its contemporary context.

The study of iconography is not limited to the study of artists of the past. Scholars who write about the Mexican painter Frida Kahlo (1907–54), for instance, call attention to her use of Christian, Aztec, and Marxist images. They point out that in her *Self-Portrait with Thorn Necklace and Hummingbird* (1940), the thorn necklace alludes not only to the crown of thorns placed on Jesus's head—Kahlo as guiltless sufferer—but probably also alludes to the mutilation that Aztec priests inflicted on themselves with thorns and spines. The hummingbird, sacred to Huitzilopichtli, god of the sun and of war, is an Aztec symbol of the souls of warriors who died in battle or on a sacrificial stone. Or take another of Kahlo's self-portraits, *Marxism Will Give Health to the Sick* (1954), painted a few months before she died (see page 168). As noted earlier in this chapter, in her youth Kahlo had been partly paralyzed by a traffic accident. In this picture she shows herself strapped in an orthopedic corset but discarding her crutches although one of her legs had been amputated in the year before she painted this picture. (The painting is in the tradition of an *ex-voto*, a picture given to a church in fulfillment of a vow made by someone who prayed to a saint for help, and whose prayer was answered, for instance, by a miraculous cure. Such pictures customarily show the saint at the top and the recipient of the miracle in the center; for Kahlo, Karl Marx is contemporary humanity's savior.) In the upper right, Marx strangles an eagle with Uncle Sam's head (i.e., Marxism destroys American imperialistic capitalism). Below Uncle Sam are red rivers—presumably rivers of the blood of America's victims—and a shape that probably represents the mushroom cloud of an atomic bomb. At the top left, the dove of peace counterbalances the wicked eagle. Beneath the dove a globe shows Russia, from which flow not rivers of blood but blue rivers, rivers of life-giving, cleansing water. Kahlo, holding a red book (Marx's teachings), is supported by large hands (great power) near Marx; in one of the hands is an eye, a symbol of knowledge (Marx sees all and understands all).

The Tehuana dress that Kahlo wears in this picture also appears in several of her other paintings. Janice Helland, speaking of this dress in another painting, explains the iconography:

> This traditional costume of Zapotec women from the Isthmus of Tehuantepec is one of the few recurring indigenous representations in

Frida Kahlo, *Marxism Will Give Health to the Sick*, 1954. Oil on masonite, 30″ × 24″. (Collection of the Frida Kahlo Museum, Mexico City. Photograph: Cenidiap/INBA, Mexico. Reproduction authorized by the Instituto Nacional de Bellas Artes y Literatura/Courtesy Banco de México, Fiduciario en el Fideicomiso Relativo a Los Museos, Diego Rivera y Frida Kahlo)

Kahlo's work that is not Aztec. Because Zapotec women represent an ideal of freedom and economic independence, their dress probably appealed to Kahlo.

—"Aztec Imagery in Frida Kahlo's Paintings," *Women's Art Journal* 11 (Fall 1990–Winter 1991): 9–10

Helland cites three references supporting her interpretation of the Tehuana dress.

Iconology (Greek for "image study") seeks to relate the symbolic meanings of objects and figures in art to (in the words of Erwin Panofsky) "the political, poetical, religious, philosophical, and social tendencies of the personality, period or country under investigation" (*Studies in Iconology,* 1939, reprinted 1967, page 16). That is, iconology interprets an image for evidence of the cultural attitudes that produced what can be called the meaning or content of the work. For instance, iconology can teach us the significance of changes in pictures of the Annunciation, in which the angel Gabriel confronts Mary. These changes reveal cultural changes. Early paintings show a majestic Gabriel and a submissive Virgin. Gabriel, crowned and holding a scepter, is the emblem of sovereignty. But from the fifteenth century onward the Virgin is shown as Queen of the Angels, and Gabriel, kneeling and thus no longer dominant, carries a lily or a scepter tipped with a lily, emblem of the Virgin's purity. In this example, iconology—the study of iconography—calls to our attention evidence of a great change in religious belief. In short, iconology tells us *why* images mean what they mean. (Panofsky, who introduced the terms, later dropped "iconology" and preferred to speak only of "iconography" and "iconographical interpretation.")

The identification of images with symbolic content is not, of course, limited to images in Western art. Here is a brief passage discussing a veranda post for a palace, carved by an African sculptor, Olowe (c. 1875–1938), whom John Pemberton III calls "perhaps the greatest Yoruba carver of the twentieth century." The post (see page 170), shows a seated king wearing a conical beaded crown that is topped by a bird whose beak reaches halfway down the crown. Beneath the king are a kneeling woman and a palace servant, and behind the king is the senior queen. Pemberton says:

> When the crown . . . is placed upon his head by the senior queen, his destiny (*ori*) is linked to all who have worn the crown before him. The great bird on the crown refers to "the mothers," a collective term for female ancestors, female deities and for older living women, whose power over the reproductive capacities of all women is held in awe by Yoruba men. Referring to the cluster of birds on his great crown, the Orangun-Ila said: "Without 'the mothers,' I could not rule." Thus, the bird on the Ogoga's crown and the senior queen, whose breasts frame the crown, represent one and the same power—the hidden, covert, reproductive power of women, upon which the overt power of Yoruba kings ultimately depends. . . .
>
> —John Pemberton III, "The Carvers of the Northeast," in *Yoruba: Nine Centuries of African Art and Thought,* ed. Allen Wardwell (New York: Center for African Art, 1989), 206

Until fairly recently, discussions of African art rarely went beyond speaking of its "brute force," its "extreme simplifications," and its influence on Picasso and other European artists. But it is now recognized that African

The central veranda post carved by Olowe of Ise for the palace courtyard of the Ogaga of Ikere, Nigeria. (Olowe of Ise, d. 1938, Africa, Nigeria, Ekiti, Ikere; Yoruba, Veranda post of Enthroned Kin [Opo Ogoga], wood, pigment, 1910/14, 152.5 × 31.7 × 40.6 cm. Major Acquisitions Centennial Fund, 1984.550. Right 3/4 view. Photograph by Bob Hashimoto, photograph © 1999 The Art Institute of Chicago. All rights reserved.

art, like the art of other cultures, expresses thought—for instance, ideas of power such as prestige, wealth, and fertility—in material form. In short, the forms of African art embody world views. (See, for example, Suzanne Preston Blier, *The Royal Arts of Africa*, 1998.)

This discussion of iconography has spoken of "the proper identification of an image." Here we have a clue to the chief assumption held by most people who study iconography: A work of art is a unified whole, and its meaning is what the creator took it to be or intended it to be. In our discussion of meaning (pages 16–21), however, we saw that many art historians today (especially those associated with the New Art History) would argue with this assumption.*

---

*For a discussion of the strengths and limitations of iconographic studies, see the excellent introduction to Brendan Cassidy, ed., *Iconography at the Crossroads* (Princeton: Index of Christian Art, 1993). See also Roelof van Straten, *An Introduction to Iconography* (Langhorne, Pa.: Gordon and Breach, 1994).

7

Art-Historical Research

It is sometimes argued that there is a clear distinction between *scholarship* (or *art-historical research*) and *criticism*. In this view, scholarship gives us information about works of art and it uses works of art to enable us to understand the thought of a period; criticism gives us information about the critic's feelings, especially the critic's evaluation of the work of art. Art history, it has been said, is chiefly fact-finding, whereas art criticism is chiefly fault-finding. And there is some debate about which activity is the more worthwhile. The historical scholar may deprecate evaluative criticism as mere talk about feelings, and the art critic may deprecate scholarly art-historical writing as mere irrelevant information. But before we further consider the relationship between historical scholarship and criticism, we should think briefly about a third kind of activity, *connoisseurship*.

Connoisseurship

The connoisseur identifies and evaluates works of art. Erwin Panofsky, in *Meaning in the Visual Arts* (Garden City, NY: Doubleday, 1955), suggests that the connoisseur differs from the art historian not so much in principle as in emphasis; the connoisseur's opinions (e.g., "Rembrandt around 1650"), like the historian's, are verifiable. The difference (Panofsky says) is this: "The connoisseur might be defined as a laconic art historian, and the art historian as a loquacious connoisseur" (page 20). But elsewhere in his book Panofsky distinguishes between art history on the one hand and, on the other, "aesthetics, criticism, connoisseurship, and 'appreciation'" (page 322). Perhaps we can retain Panofsky's definition of the connoisseur as a "laconic art historian" and say that the connoisseur's specialty is a sensitivity to artistic traits. We can say, too, that the art historian possesses the knowledge of the connoisseur—a knowledge of what is genuine and of when it was made—and then goes on (by analyzing forms and by relating them chronologically) to explain the changes that have occurred in the ways that artists have seen.

Connoisseurship today has been widely censured, especially by left-ist historical scholars who sometimes characterize themselves as practi-tioners of the New Art History (see pages 151–56). These scholars, con-cerned largely with art as a revelation of the social and political culture that produced it, see connoisseurship as an arid activity that chiefly serves rich people. In the eyes of the New Art History, connoisseurship concentrates on minutiae, evades confronting the social implications of art, fosters the elitist implications of art museums, bolsters the art market by authenticating works for art dealers and collectors, and asserts the va-lidity of such mystical and elitist concepts as "taste" and "intuition" and "quality."

Even a new sort of connoisseurship, relying on scientific tests of pig-ments, paper, patination, and so forth, has been criticized along the same grounds. Scientific study is intended to clear matters up, but persons skeptical of connoisseurship argue that highly technical reports serve chiefly to enhance the mystique of art. On the other hand, many connois-seurs (they are found chiefly in museums and in dealer's shops rather than in colleges and universities) believe that art historians tend to be in-sufficiently concerned with works of art as things of inherent worth and overly concerned with art as material for the study of political or intellec-tual history. But let us now further consider the relationship between the art historian and the art critic.

History and Criticism

The *art historian* is sometimes viewed as a sort of social scientist, recon-structing the conditions and attitudes of the past through documents. (The documents, of course, include works of art as well as writings.) In studying Cubism, for example, the supposedly dispassionate art historian does not prefer one work by Braque to another by Braque, or Picasso to Braque, or the other way around. The historian's job, according to this view, is to explain how and why Cubism came into being, and value judg-ments are considered irrelevant.

The *art critic,* on the other hand, is supposedly concerned not with verifiable facts but with value judgments. Sometimes these judgments can be reduced to statements such as "This work by Braque is better than that work by Picasso," or "Picasso's late works show a falling-off," and so on, but even when critics are not so crudely awarding As and Bs and Cs, acts of evaluation lie behind their choice of works to discuss. Intrigued by a work, they usefully call our attention to qualities in it that evoke a response, helping us to see what the work has to offer us. That is, (1) they offer a considered response, revealing elements in the work so that (2) we can better experience the work. In the words of the novelist D. H. Lawrence, criticism offers "a reasoned account of the feelings" produced by a work.

SOME CRITICAL VALUES

Before we consider standards of evaluation, let's pause to think about evaluation in general. When we say, "This is a great picture," are we in effect saying only "I like this picture"? Are we merely *expressing our taste* rather than *pointing to something out there,* something independent of our tastes and feelings? Consider three sentences:

1. It's raining outside.
2. I like vanilla.
3. Of Cézanne's three versions of *The Card Players,* the one in the Louvre is the best.

If you are indoors and you say that it is raining outside, a hearer may ask for verification of this assertion about something outside. Why do you say what you said? "Because Jane just came in, and she's drenched," or "Because I just looked out of the window." If, on the other hand, you say that you like vanilla, it's almost unthinkable that anyone would ask you why. No one expects you to justify—to offer a reason in support of—an expression of taste.

   Now consider the third statement, that the Louvre version of *The Card Players* is the best of the three paintings. (This assertion is made by Meyer Schapiro, in his *Paul Cézanne,* 1952, on page 88.) Does this assertion resemble the assertion about the weather, or does it resemble the assertion about vanilla? The weather, because it would have been entirely reasonable for someone to ask Schapiro *why* he said the Louvre version is the best. And in fact in his book Schapiro goes on to give his reasons. He says the Louvre version is the most monumental and the most varied, and he supports these assertions by pointing to particular details. His statement evaluating the three pictures asserts something that we can discuss, in a way that we cannot discuss the expression of a personal preference for vanilla. We can, so to speak, hold a mental conversation with Schapiro, offering our own views in response to his. We might begin by thinking about *why* he might value monumentality and variety. We may conclude that Schapiro's values are merely subjective; or we may conclude (at the opposite pole) that these qualities indeed are present in all the works that are valued by virtually all persons who think carefully about the issue. Or (a middle position) we may conclude that although certain qualities are widely esteemed, it becomes evident that it is only certain groups in certain historical periods that esteem them. (This last view is called sociohistorical relativism.) These are issues that can be discussed.

Consider this passage about Picasso's *Les Demoiselles d'Avignon* (see page 25 for a reproduction):

> *The Young Ladies of Avignon,* that great canvas which has been so frequently described and interpreted, is of prime importance in the sense of being the concrete outcome of an original vision, and because it points to a radical change in the aesthetic basis as well as the technical processes of painting. In itself the work does not bear very close scrutiny, for the drawing is hasty, and the colour unpleasant, while the composition as a whole is confused and there is too much concern for effect and far too much gesticulation in the figures. . . . The truth is that this famous canvas was significant for what it anticipated rather than for what it achieved.
>
> —Frank Elgar and Robert Maillard, *Picasso,* trans. Francis Scarfe (1956), 56–58; quoted in Monroe Beardsley, *Aesthetics* (New York: Harcourt, 1958), 454

Notice that Elgar and Maillard evaluate the painting on at least two very different grounds: (1) historical significance (the picture is important because it is "original" and it altered the history of painting) and (2) aesthetic merit or inherent worth (the picture is not very important because "the drawing is hasty," the color is "unpleasant," "the composition . . . is confused," "there is too much concern for effect," and there is "far too much gesticulation in the figures"). Probably if they were pressed they would go on to explain—by pointing to details—what they meant by saying that the drawing was "hasty," the color "unpleasant," and so forth. Certainly a reader feels that the authors *ought* to be able to support their views or they shouldn't have asserted them. In our mental conversation with Elgar and Maillard we might begin by asking *why* hastiness is a weakness? Aren't there drawings and paintings that impress us because they seem spontaneous? Conversely, aren't there highly finished works that we censure because they seem fussy or labored? In short, *is* evidence

Frank Lloyd Wright, *Falling Water,* the Edgar J. Kaufmann House, 1936–38. Bear Run, Pennsylvania. The house is cantilevered out over a stream. (Western Pennsylvania Conservancy/Art Resource, NY)

of hastiness a weakness and *is* evidence of great care a virtue? That's worth thinking about. And what about "gesticulation"? As we think about this critical comment, we may conclude that the writers have not adequately supported their assertions.

Let's look at another passage of evaluative writing, a paragraph in which the architect Paul M. Rudolph talks about Falling Water or Fallingwater (1936–38), a house Frank Lloyd Wright designed as the weekend retreat for a department store magnate:

> Fallingwater is that rare work which is composed of such delicate balancing of forces and counterforces, transformed into spaces thrusting

horizontally, vertically and diagonally, that the whole achieves the serenity which marks all great works of art. This calmness, with its underlying tensions, forces and counterforces, permeates the whole, inside and out, including the furnishings and fittings. . . . The thrusts are under complete control, resulting in the paradox of a building full of movement: turning, twisting, quivering movement—which is, at the same time, calm, majestic and everlasting.

—"Fallingwater," *Global Architecture* 2 (1970)

Rudolph, in love with "underlying tensions, forces and counterforces," is barely concerned with how well Falling Water works as a house. Is it noisy? (Yes.) Damp and hard to heat? (Yes.) Can one see and enjoy the waterfall from inside? (No.) Rudolph does not ask these questions because he does not care about how the house functions. In fact, he goes on to suggest that "the mark of an architect is how he handles those spaces that are not strictly functional." In short, Rudolph's values (at least in this essay) have to do chiefly with the resolution of forces, little to do with functionalism, and nothing to do with economics or politics. Another critic—the client, for instance—might evaluate the house in terms of livability. (Le Corbusier said, "A house is a machine for living in." How well does this machine work?) And still another critic—a Marxist, perhaps—might evaluate the building as a ridiculously expensive toy designed for a millionaire. These and other critics would come up with a far less favorable evaluation.

In the preceding pages we saw critics setting forth criteria: Schapiro valued monumentality and variety in Cézanne's *The Card Players,* Elgar and Maillard regarded hastiness as a fault, Rudolph valued "underlying tensions" but also said that "serenity . . . marks all great works of art." Let's take a brief survey of some theories of value that have been common in discussions of Western art.°

°Rudolph's mention of "serenity" is significant because a good deal of Western writing about art focuses on pleasure. But some other cultures produce objects in accordance with other criteria. Suzanne Preston Blier, in *African Vodun: Art, Psychology, and Power* (Chicago: University of Chicago Press, 1995) quotes a Fon-speaking West African sculptor who says, "One makes things that will cause fear" (page 59). For introductory material on non-Western topics such as "Eskimo Aesthetics," "Navajo Aesthetics," "Aztec Aesthetics," and "Yoruba Aesthetics," see Richard L. Anderson, *Calliope's Sisters: A Comparative Study of Philosophies of Art* (Englewood Cliffs, N.J.: Prentice Hall, 1990). There has been little study of such topics, and the findings are deeply contested, but Anderson's bibliography is valuable.

Truth. One view holds that art should be true. At its simplest, perhaps, this means that it should be realistic, presenting images that closely resemble what we see when we look at the world around us. This is what the novelist Emile Zola implied in 1866 when he wrote of Corot's paintings, "If M. Corot would kill, once and for all, the nymphs of his woods and replace them with peasants, I should like him beyond measure." The idea that truth confers value on art is by no means dead. Thus, Allan Sekula, a critic with a Marxist approach, wants artists to present important economic and social truths. He finds inadequate the liberal sentiments usually associated with documentary photographers such as W. Eugene Smith. What Sekula wants is a politically sophisticated art (in this case, a documentary photograph) that reveals to the viewer what Sekula takes to be the truth about capitalism:

> For all his good intentions, for example, Eugene Smith in *Minamata* provided more a representation of his compassion for mercury-poisoned fisherfolk than one of their struggle for retribution against the corporate polluter. I will say it again: the subjective aspect of liberal aesthetics is compassion rather than collective struggle. Pity, mediated by an appreciation of "great art," supplants political understanding.
>
> —Allan Sekula, *Photography Against the Grain* (Halifax: Nova Scotia College of Art and Design, 1984), 67

For Sekula it is not enough for a picture to have formal beauty (say, the "underlying tensions" that Paul Rudolph valued in Falling Water) or for a picture to evoke compassion. It must also reveal and convey political understanding. "I am arguing, then," Sekula says, "for an art that documents monopoly capitalism's inability to deliver the conditions of a fully human life" (page 74).

In talking about art and truth—whether we are holding the work of art up against the truth as it is given by a religious system (for instance, Christianity) or by a secular system (for instance, Marxism)—we are saying that the value of the work depends on something outside of the work itself, something *extrinsic* to the work, here the external world, seen through Christian eyes or seen through Marxist eyes. These critics say that a work of art is good if it has an **instrumental value;** that is, the work is an *instrument,* improving us spiritually or morally, or giving us insight into the political system so that we can work for a more just system.

Intrinsic Aesthetic Merit. In contrast to value systems that judge the work in terms of something extrinsic to the work—such as the real world, as understood by Christians or by Marxists, or the artist's sincerity or insincerity, as conjectured by the critic—we can place **formalism.** Formalism holds that *intrinsic* qualities—qualities within the work, having no reference to the outside world—make it good or bad. We have seen Paul Rudolph praise Falling Water because of its "delicate balancing of forces and counterforces" and because "the whole achieves the serenity which marks all great works of art."

Are there patterns that are inherently pleasing, patterns that in themselves afford aesthetic pleasure to the viewer? Euclid, the Greek mathematician of the third century BC, defined one pattern that has come to be called *the golden section,* a pattern much discussed in the Renaissance—for instance, by Leonardo—and unquestionably used by later artists such as Claude Lorrain (1600–82) and Richard Wilson (1714–82). A line or rectangle is divided into two unequal parts, so that the ratio of the smaller is to the larger as the ratio of the larger is to the whole. This proportion, roughly, is 3 to 5; thus (again, roughly), 3 is to 5 as 5 is to 8. In a landscape, the painter might divide the canvas vertically, giving us classical ruins in the smaller portion and an expansive landscape in the larger portion. Or the line might be conceived horizontally, dividing the canvas into an upper and a lower portion. In a landscape painting the horizon might be drawn so that the sky occupied the larger rectangle, the land the smaller. Whether the golden section does offer aesthetic pleasure is still a matter of controversy—and if it does, does it do so for people of all cultures?—but the point here is merely to explain the idea that in the formalist view, a work offers pleasure by means of its form. If you have ever rearranged objects on a mantel, or even straightened a picture on a wall, you know that some patterns are (at least for you) more pleasing than others.

One additional point: Although most formalist critics suggest that the work is not to be judged by anything outside of itself, some formalists do hold an instrumental position, claiming that works of art are valuable because by means of their harmony they induce mental harmony in the viewer. "Music," William Congreve said (but perhaps we can extend his remark to any work of art), "has charms to soothe a savage breast," and indeed some racetracks pipe music into the stables to soothe the highstrung horses.

Modernist critics, active chiefly from about 1950 to the mid-1970s, were primarily committed to abstract art and the apparently impersonal no-frills architecture of the International Style epitomized in the work of Mies van der Rohe. They tended to assume, as formalist critics, that a work of art is self-sufficient, pure, the product of a genius and therefore opposed to popular culture, and it need not—indeed should not—be seen as a reflection of the social context. It now seems evident, however, that admirers of the International Style were espousing—whether they knew it or not—a capitalist celebration of business efficiency. For the modernist, the significant artist was highly original, a member of the avant-garde, a genius who produced a unique work that marked an advance in the history of art—a sort of artistic Thomas Edison or Henry Ford. Not surprisingly, modernist critics tended to practice formal analysis.

A reaction against the modernist critics of the mid-twentieth century probably was inevitable, and it took the name of **post-modernism.** Post-modernist critics (active from about 1970 to the present) argue that the supposedly dispassionate old-style art historians are, consciously or not, committed to the false elitist ideas that universal aesthetic criteria exist and that only certain superior things qualify as "art." For post-modernists, modernist art and modernist criticism are all dressed up with nowhere to go. Here is the way Tom Gretton puts it:

> For most art historians "art" does not designate a set of types of object—all paintings, sculptures, prints and so forth—but a subset arrived at by a more or less openly acknowledged selection on the basis of aesthetic criteria. But aesthetic criteria have no existence outside a specific historical situation; aesthetic values are falsely taken to be timeless.
> —Tom Gretton, "New Lamps for Old," in *The New Art History*, ed.
> A. L. Rees and Francis Borzello (London: Camden Press, 1988), 64

Behind a good deal of post-modernist criticism stand the writings of the architects Robert Venturi and Denise Scott-Browne, who see modernism, with its emphasis on formal beauty, as narcissistic. Post-modern critics, seeing the artist as deeply implicated in society, reject formal analysis and tend to discuss artworks not as beautiful objects produced by unique sensibilities but as works that exemplify a society's culture, especially its politics. This approach has already been discussed (see pages 84, 151–56) in comments on deconstruction and on the New Art History.

Other Criteria. We have not exhausted the criteria that critics may offer. Some critics value art for its **expressiveness.** In this view, artists express themselves, setting forth their inner life. In Jackson Pollock's words, "Painting is a state of being . . . self-discovery. Every good artist paints what he is." It may be healthful for artists to express themselves, but what is the value for viewers? The usual answer is that the work of art gives viewers (if the expression is successful) new states of feeling. Thus, Jonathan Fineberg in *Art Since 1940* (1995) says of Eva Hesse—for an example of her work, see page 58—"In Hesse's case, the sheer force of her drive to find form (in whatever material seemed most evocative) for her profound emotional struggles presses the viewer irresistibly into identifying with her discovery of herself in the objects she created" (page 311).

Critics who value art for its expressiveness are likely to talk about the artist's **sincerity,** faithfulness to his or her vision, and so on. Skeptics, however, may ask how we can know if the artist indeed was sincere and faithful. Do we turn to the artist's letters, for instance, or to statements made in interviews, in order to validate the sincerity of the work? How can we confirm Fineberg's assertion (pages 16–17) that "Courbet painted with *so much conviction* that over time his way of seeing won over more and more artists and observers to his point of view" (italics added)? True, we can turn to documents if we are talking about modern artists, and we can agree that much of the interest in modern art is in the artist's highly personal response to experience, but what about older artists who have left us little or nothing besides the works themselves? How do we know if the cave paintings at Lascaux, the sculptures from the Parthenon, and *Mona Lisa* are sincere? And can't a very bad artist be utterly sincere?

Still other critics may especially value **technique.** We have only to look at the perfection of the potting of, say, a Chinese porcelain—the uniformity of the color, the almost unbelievable thinness of the clay—to see value in the work. The commonly heard complaint, "Any two-year-old could do that," is a complaint about the apparent lack of skill. In valuing technique (skillful execution) we may, when speaking of paintings, be getting back to realism: "Look how the painter has caught the texture of fur here, and the texture of silk there. And just look at the light reflecting on that glass of wine!"

Originality is yet another value. The first artist to do something gets extra credit, so to speak, even if on other grounds the work is not exceptional. Here we recall Elgar and Maillard's praise (page 175) of Picasso's *Les Demoiselles d'Avignon* as "the concrete embodiment of an original vision." In this vein, the sculptor Richard Serra has said (*The New York Times,* 11 August 1995, C1) that he inclines to "evaluate artists by how much they are able to rid themselves of convention, to change history." Monet's late paintings of water lilies were for several decades regarded as of no interest and were unsaleable, but fairly recently they have been hailed as precursors of Abstract Expressionism and their prices have soared. Conversely, an artist who pretty much keeps doing the same thing may be put down as unimaginative, even though each work may be excellent when judged by other criteria.

Even this very brief survey of conflicting values offers enough to indicate why some philosophers doubt that objective evaluation is possible. And one can see why they say that such terms as "beautiful" and "ugly" are not really statements about artworks but are simply expressions of feelings, mere emotional responses, high-class grunts of approval or sighs of despair. (However, as we saw on pages 17–19, some philosophers of art insist that the real existence of a work is indeed in the observer's response, not in its material makeup of paint on canvas or a mass of bronze.)

Probably most people today would agree that there are no inflexible rules by which we can unerringly judge all works. Still, the preceding paragraphs seek to convince you that discussions of artistic value are not merely subjective, not just expressions of personal opinion. When we are deeply moved by an artwork, we want others to feel equally excited, and so we point to details, we offer *arguments* to convey *why* we care about the work. Sometimes others agree with us immediately; sometimes we have to offer further analysis; and sometimes, no matter how precise and passionate our arguments, we fail to convince our audience. Then it becomes their turn, and the discussion and debate proceed. Our own responses may sharpen or even radically change. Fine; this is part of why art, and critical conversations about art, can be richly rewarding. But good critics recognize a duty to set forth their feeling in (to repeat D. H. Lawrence's words) a *reasoned* account.

If you want some fun while thinking about what makes a work of art good or bad, visit the Web site of the Museum of Bad Art, *www.glyphs.com/moba*, or visit the museum itself, in the concrete basement (next to the men's room) of the community theater in Dedham, Massachusetts.

✍ A RULE FOR WRITERS:

When you draft an essay, and when you revise it through successive drafts, imagine that you are explaining your position to someone who, quite reasonably, wants to hear the *reasons* that have led you to your conclusions.

HISTORICAL SCHOLARSHIP AND VALUES

The root of much historical writing, like that of critical writing, is also a feeling or intuition—a hunch perhaps that a stained glass window in a medieval cathedral may be a late replacement, or that photography did not influence the paintings of Degas as greatly as is usually thought. The historian then follows up the hunch by scrutinizing the documents and by setting forth—like the critic—a reasoned account.

It is doubtful, then, that historical scholarship and aesthetic criticism are indeed two separate activities. To put it a little differently, we can ask if scholarship is really concerned exclusively with verifiable facts, or, on the other hand, if criticism is really concerned exclusively with unverifiable responses (i.e., only with opinions). If scholarship limited itself to verifiable facts, it would have little to deal with; the verifiable facts usually don't go far enough.

Suppose that a historian who is compiling a catalog of Rembrandt's work, or who is writing a history of Rembrandt's development, is confronted by a drawing attributed to Rembrandt. External documentation is lacking: No letter describes the drawing, gives a date, or tells us that the writer saw Rembrandt produce it. The historian must decide whether the drawing is by Rembrandt, by a member of Rembrandt's studio, by a pupil but with a few touches added by the master, or is perhaps an old copy or even a modern forgery. Scholarship (e.g., a knowledge of paper and ink) may reveal that the drawing is undoubtedly old, but questions still remain: Is it by the master or by the pupil or by both?

Even the most scrupulous historians must bring their critical sense into play and offer conclusions that go beyond the verifiable facts—conclusions that are ultimately based on an evaluation of the work's quality, a feeling that the work is or is not by Rembrandt and—if the feeling is that the work *is* by Rembrandt—a sense of *where* it belongs in Rembrandt's chronology. Art historians try to work with scientific objectivity, but because the facts are often inconclusive, much in their writing is inevitably (and properly) an articulation of a response, a rational explanation of feeling that is based on a vast accumulation of experience. Another example: A historian's decision to include in a textbook a discussion of a given artist or school of art is probably a judgment based ultimately on the feeling that the matter is or is not worth discussing, is or is not something of importance or value. And the decision to give Vermeer more space than Casper Netscher is an aesthetic decision, for Netscher was, in his day, more influential than Vermeer. Indeed, art history has worked along these lines from its beginnings in *Lives of the Most Excellent Painters, Sculptors and Architects* by Giorgio Vasari (1511–74). Vasari says that he disdained to write a "mere catalog," and that he did not hesitate to "include [his] own opinion" everywhere and to distinguish among the good, the better, and the best.

On the other hand, even critics who claim that they are concerned only with evaluation, and who dismiss all other writing on art as sociology or psychology or gossip, bring some historical sense to their work. The exhibitions that they see have usually been organized by scholars, and the accompanying catalogs are often significant scholarly works; it is virtually impossible for a serious museum-goer not to be influenced, perhaps unconsciously, by art historians. When, for example, critics praise the Cubists for continuing the explorations of Cézanne and damn John Singer Sargent for contributing nothing to the development of art, they are doing much more than expressing opinions; they are drawing on their knowledge of art history, and they are echoing the fashionable view that a work of art is good if it marks an advance in the direction that art happens to have taken. (One can in fact point out that there are so many trends, one can doubt that art is going in any particular direction.)

Probably it is best, then, not to insist on the distinction between scholarship and criticism, but to recognize that most writing about art is a blend of both. True, sometimes a piece of writing emphasizes the facts that surround the work (e.g., sources, or the demands of the market), showing us how to understand what people once thought or felt; and sometimes it emphasizes the reasons why the writer especially values a particular work, showing the work's beauty and significance for us. But on the whole the best writers on art do both things, and they often do them simultaneously.

Consider, as an example, Julius S. Held's article, "The Case Against the Cardiff 'Rubens' Cartoons." Four full-scale cartoons (large drawings for transfer) for tapestries were acquired in 1979 by the National Museum of Wales as works by Rubens, but their attribution has been doubted. In the second paragraph of his article, Professor Held comments:

> When I first saw the photographs of the cartoons, and even more so after I had seen them in their setting in Cardiff (27th July 1980), I failed to notice anything that would justify the extraordinary claims made for them. That they were Flemish paintings of the seventeenth century no one could deny. They could also be of special interest as cartoons painted on paper, a category of works that must have existed in great

quantity, though few have survived. Yet an attribution to Rubens, as the author of their design as well as the actual executant seemed to me highly questionable and certainly not at all self-evident.

—Julius S. Held, "The Case Against the Cardiff 'Rubens' Cartoons," *Burlington Magazine* (March 1983), 132

Held is suggesting that the cartoons do not look like works by Rubens, and in the words "the extraordinary claims made for them" we hear a strong implication that they are not good enough to be by Rubens. So we begin with evaluation, taste, perhaps subjective judgment, but judgment made by a trained eye.

Held goes on to buttress this judgment with other criteria, of a more objective kind. Technically, the cartoons are unprecedented in Rubens's oeuvre. They are done in watercolor on paper, a medium he never used on this scale; even in small drawings watercolor alone is rarely employed. The attribution, hence, of the disputed cartoons to Rubens is unlikely on this ground alone. Held proceeds to ask himself on what basis these cartoons were attributed to Rubens in the first place. The main argument in favor was based on the existence of two *modelli,* painted in oil and depicting actions similar to those of the two cartoons. (A *modello*—the plural is *modelli*—is a small version shown to a patron or prospective patron before the large picture is executed.) These *modelli* are indeed by Rubens, but they cannot be used in support of the attribution of the cartoons. Why? Because in composition these *modelli* differ radically from the cartoons, whereas in all the instances in which we know Rubens's *modelli* for tapestry cartoons, the *modelli* and the cartoons agree with each other in all essential respects. In disengaging the *modelli* from the cartoons, Held also uses an iconographic argument: The cartoons depict incidents from the life of Aeneas, but the *modelli* illustrate scenes from the life of Romulus.

In short, during the course of his argument, Held introduces historical data in support of critical judgments. His final words (page 136) make this clear: "The purpose of my paper is served . . . if it succeeds to remove Rubens's name from four paintings that are not worthy to carry it."

8

Writing a Research Paper

Because a research paper requires its writer to collect the available evidence—usually including the opinions of earlier investigators—one sometimes hears that a research paper, unlike a critical essay, is not the expression of personal opinion. But such a view is unjust both to criticism and to research. A critical essay is not a mere expression of personal opinion; to be any good it must offer evidence that supports the opinions, thus persuading the reader of their objective rightness. And a research paper is largely personal because the author continuously uses his or her own judgment to evaluate the evidence, deciding what is relevant and convincing. A research paper is not merely an elaborately footnoted presentation of what a dozen scholars have already said about a topic; it is a thoughtful evaluation of the available evidence, and so it is, finally, an expression of what the author thinks the evidence adds up to.°

You may want to do some research even for a paper that is primarily critical. Consider the difference between a paper on the history of Rodin's reputation and a paper offering a formal analysis of a single work by Rodin. The first of these, necessarily a research paper, will require you to dig into books and magazines and newspapers to find out about the early response to his work; but even if you are writing a formal analysis of a single piece, you may want to do a little research into, for example, the source of the pose. The point is that writers must learn to use source material thoughtfully, whether they expect to work with few sources or with many.

Although research sometimes requires one to read tedious material, or material that, however interesting, proves to be irrelevant, those who engage in research feel, at least at times, an exhilaration, a sense of triumph at having studied a problem thoroughly and at having arrived at

---

°Because footnotes may be useful or necessary in a piece of writing that is *not* a research paper (such as this chapter), and because I want to emphasize the fact that a thoughtful research paper requires more than footnotes, I have put the discussion of footnotes in Chapter 9, "Manuscript Form" (pages 221–64).

conclusions that at least for the moment seem objective and irrefutable. Later, perhaps, new evidence will turn up that will require a new conclusion, but until that time, one may reasonably feel that one knows *something*.

PRIMARY AND SECONDARY MATERIALS

The materials of most research are conventionally divided into two sorts, primary and secondary. The *primary* materials or sources are the subject of study, the *secondary* materials are critical and historical accounts already written about these primary materials. For example, if you want to know whether Rodin was influenced by Michelangelo, you will look at works by both sculptors, and you will read what Rodin said about his work as a sculptor. In addition to studying these primary sources, you will also read secondary material such as modern books on Rodin. There is also material in a sort of middle ground: what Rodin's friends and assistants said about him. If these remarks are thought to be of compelling value, especially because they were made during Rodin's lifetime or soon after his death, they can probably be considered primary materials. And for a work such as Rodin's *Monument to Balzac* (1897), the novels of Balzac can be considered primary materials.

If possible, draw as heavily as you can on primary sources. If in a secondary source you encounter a quotation from Leonardo or Mary Cassatt or whomever—many artists wrote a good deal about their work—do not be satisfied with this quotation. Check the original source (it will probably be cited in the secondary source that quotes the passage) and study the quotation in its original context. You may learn, for instance, that the comment was made so many years after the artwork that its relevance is minimal.

FROM SUBJECT TO THESIS

First, a subject. No subject is undesirable. As G. K. Chesterton said, "There is no such thing on earth as an uninteresting subject; the only thing that can exist is an uninterested person." Research can be done on almost anything that interests you, though you should keep in mind two limitations. First, materials for research on recent works may be extremely difficult to get hold of, since crucial documents may not yet be in print and you may not have access to the people involved. Second, materials on some subjects may be unavailable to you because they are in languages you can't read or in publications that no nearby library has. So you probably won't try to work on the stuff of today's news—for example, the legal disposition of the works of a sculptor whose will is now being contested; and (because almost nothing in English has been written on it) you won't try to work on the date of the introduction into Japan of the image of the Buddha at birth. But no subject is too trivial for study: Newton, according to legend, wondered why an apple fell to the ground.

You cannot, however, write a research paper on subjects as broad as Buddhist art, Michelangelo, or the Asian influence on Western art. You have to focus on a much smaller area within such a subject, and you will have to have a *thesis,* a point, a controlling idea, that you will be making. Rebecca Bedell's essay on the American painter John Singleton Copley is not simply a survey of Copley's work or a comparison of two portraits, but an argument, that is, a developed presentation of the evidence supporting a thesis. In her second paragraph (page 108) Bedell states the thesis:

> Copley reached his artistic maturity years before he left for England.

The essay as a whole argues on behalf of this point.

Suppose you are interested in the Asian influence on Western art. You might narrow your topic so that you concentrate on the influence of Japanese prints on van Gogh or Mary Cassatt or Whistler (it may come as a shock to learn that the famous picture commonly called "Whistler's Mother" is indebted to Japanese prints), or on the influence of calligraphy on Mark Tobey, or on the influence of Buddhist sculpture on Jo Davidson. Your own interests will guide you to the topic—the part of the broad subject—that you wish to explore, and you won't know what you wish to explore until you start exploring. Picasso has a relevant comment: "To know what you want to draw, you have to begin drawing. If it turns out to be a man, I draw a man."

Of course, even though you find you are developing an interest in an appropriately narrow topic, you don't know a great deal about it; that's one of the reasons you are going to do research on it. Let's say that you happen to have a Japanese print at home, and your instructor's brief reference to van Gogh's debt to Japanese prints has piqued your interest. You may want to study some pictures and do some reading now. As an art historian (at least for a few hours each day for the next few weeks), at this stage you think you want to understand why van Gogh turned to Japanese art and what the effect of Japanese art was on his own work. Possibly your interest will shift to the influence of Japan on van Gogh's friend, Gauguin, or even to the influence of Japanese prints on David Hockney in the 1970s. That's all right; follow your interests. Exactly what you will focus on, and exactly what your *thesis* will be, you may not know until you do some more reading. But how do you find the relevant material?

FINDING THE MATERIAL

You may already happen to know of some relevant material that you have been intending to read—perhaps titles listed on a bibliography distributed by your instructor—or you can find the titles of some of the chief books by looking at the bibliography in such texts as H. W. Janson's *History of Art*, fifth edition (1997), Spiro Kostof's *History of Architecture*, second edition (1995), Suzanne Preston Blier's *The Royal Arts of Africa* (New York: Abrams, 1998), Craig Clunas's *Art in China* (New York: Oxford University Press, 1997), or Penelope Mason's *History of Japanese Art* (New York: Abrams, 1993). If these books do not prove useful and you are at a loss about where to begin, consult

- the card catalog or on-line catalog of your library
- the appropriate guides to articles and book reviews in journals
- the reference books listed in the following pages

The Library Catalog: Card or Computerized°

The card catalog has cards arranged alphabetically not only by author and title but also by subject. If your library has an on-line catalog, the principle is the same but alphabetization won't matter. The catalog won't have a heading for "The Influence of Japanese Prints on van Gogh," of course, but it will have one for "Art, Japanese" (this will then be broken down into periods), and there will also be a subject heading for "Prints, Japanese," after which will be a card (or an entry in the on-line catalog) for each title that is in the library's collection. And, of course, you will look up van Gogh (who, in a card catalog, probably will be listed under Gogh), where you will find cards or entries listing books by him

---

°In the following discussion of library materials I am deeply indebted to Mary Clare Altenhofen (Fogg Art Museum Library, Harvard University) and Ruth Thomas (Mugar Memorial Library, Boston University).

(e.g., collections of his letters) and about him. Often it is useful and sometimes necessary, even for computerized catalogs where you search simply by typing keywords, to look up your subject in the *Library of Congress Subject Headings*, the five big red books that libraries keep near their catalogs.

Browsing in Encyclopedias, Books, and Book Reviews

Having checked the library catalog and written down the relevant data (author, title, call number), you begin to browse in the books, or you can postpone looking at the books until you have found some relevant articles in periodicals. For the moment, let's postpone the periodicals.

You can get an admirable introduction to almost any aspect of art by looking at a magisterial thirty-four-volume work, *The Dictionary of Art*, edited by Jane Turner (1996). *The Dictionary* contains 41,000 articles, arranged alphabetically, on artists (including 3,700 entries for architects, 9,000 for painters, and 500 for photographers), forms, materials (e.g., amber, ivory, tortoise shell, and hundreds of other materials), movements, sites, theories, and so on. The articles, which range from a few hundred words to several thousand words (e.g., 8,500 on *abstract art*), include illustrations, cross-references to other articles in *The Dictionary*, and bibliographic references (some 300,000 such references). The index (670,000 entries) will guide you to appropriate articles on your subject. It is available online as *Grove Dictionary of Art*<www.groveart.com>.

Before *The Dictionary of Art* was available, the usual starting place was *Encyclopedia of World Art*, originally a fifteen-volume work published by McGraw-Hill in 1959–68, now with two supplementary volumes (1983 and 1987) updating the bibliographies. (Don't confuse this seventeen-volume work with *McGraw-Hill Dictionary of Art*, 1969, a five-volume set containing about 15,000 entries, emphasizing biographies of artists.) *Encyclopedia of World Art* includes articles on nations, schools, artists, iconographic themes, genres, and techniques. It continues to be of some value, but for the most part it is superseded by *The Dictionary of Art*.

General encyclopedias, such as *Encyclopedia Americana* and *Encyclopaedia Britannica*, can be useful at the beginning. For certain topics, such as early Christian symbolism, *New Catholic Encyclopedia* (fifteen volumes) is especially valuable. But on this subject see also *The Oxford Companion to Christian Art and Architecture* (1996).

Let's assume that you have glanced at some entries in an encyclope-dia, or perhaps have decided that you already know as much as would be included in such an introductory work, and you now want to investigate the subject more deeply. Put a bunch of books in front of you, and choose one as an introduction. How do you choose one from half a dozen? Partly by its size—choose a fairly thin one—and partly by its quality. Roughly speaking, it should be among the more recent publications, or you should have seen it referred to (perhaps in the textbook used in your course) as a standard work on the subject. The name of the publisher is at least a rough indication of quality: A book or catalog published by a major mu-seum, or by a major university press, ought to be fairly good.

When you have found the book that you think may serve as your in-troductory study,

- read the preface to get an idea of the author's purpose and outlook
- glance at the table of contents to get an idea of the organization and coverage
- scan the final chapter or, if the book is a catalog of an exhibition, the last few pages of the introduction, where you may be lucky enough to find a summary
- look through the index, which should tell you whether the book will suit your purpose by showing you what topics are covered and how much coverage they get

If the book still seems suitable, browse in it.

At this stage it is acceptable to trust one's hunches—you are only go-ing to scan the book, not buy it or even read it—but you may want to look up some book reviews to assure yourself that the book has merit, and to be aware of other points of view. Countless reviews of art books are pub-lished not only in journals devoted to art (e.g., *Art Bulletin, Art Journal, Burlington Magazine*) but also in journals devoted to historical periods (*Renaissance Quarterly, Victorian Studies*), in newspapers and maga-zines, and, online, in *CAA.reviews*, ⟨*www.caareviews.org*⟩, since the fall of 1998. There are four especially useful indexes to book reviews:

Book Review Digest (published from 1905 onward)

Book Review Index (1965–)

Art Index (1929–)

Humanities Index (1974–)

Book Review Digest includes brief extracts from the reviews, chiefly in relatively popular (as opposed to scholarly) journals. Thus if an art book was reviewed in the *New York Times Book Review*, or in *Time* magazine, you will probably find something (listed under the author of the book) in *Book Review Digest*. Look in the volume for the year in which the book was published, or in the next year. But specialized books on art will probably be reviewed only in specialized journals on art, and these are not covered by *Book Review Digest*.

You can locate reviews by consulting **Book Review Index** (look under the name of the author of the book) or by consulting *Art Index*. (In the early volumes of **Art Index,** reviews were listed, alphabetically by the author of the review, throughout the volumes, but since 1973–74 reviews have been listed at the rear of each issue, alphabetically by the author of the book or by the title if the book has no author.) Scholarly reviews sometimes appear two, three, or even four years after the publication of the book, so for a book published in 1985 you probably will want to consult issues of *Book Review Index* and *Art Index* for as late as 1989. **Humanities Index** works the same way as *Art Index* but indexes different journals for an interdisciplinary approach. Most of these indexes are also available in electronic form.

When you have located some reviews, read them and then decide whether you want to scan the book. Of course, you cannot assume that every review is fair, but a book that on the whole gets good reviews is probably at least good enough for a start.

By quickly reading such a book (take few or no notes at this stage), you will probably get an overview of your topic, and you will see exactly what part of the topic you wish to pursue.

Indexes and Databases to Published Material

An enormous amount of material on art is published in books, magazines, and scholarly journals—too much for you to look at randomly as you begin a research project. But **indexes** and **databases** can help you sort through this vast material and locate books and articles relevant to your research topic. The most widely used indexes include:

Readers' Guide to Periodical Literature (1900–)

Art Index (1929–)

BHA: Bibliography of the History of Art (1991–)

RAA: Répertoire d'art et d'archéologie (1910–90)
RILA: International Repertory of the Literature of Art (1973–89)
ARTbibliographies MODERN (1969–)
Architectural Periodicals Index (1973–)
Avery Index to Architectural Periodicals (1934–)

These indexes also are published as computerized databases—on CD-ROMs, on the Web, and through online vendors such as DIALOG. Ask your Reference Librarian which ones are available or check your library's Web page. *Cautions:* The Web versions are regulated by leases that usually limit access to currently enrolled students. You may need a password or special software if your personal computer's Internet Service Provider (ISP) is not the same as your college's. Also, the computerized databases may cover only the most recent twenty to thirty years, so you will need to use the print indexes for earlier years.

Readers' Guide indexes more than a hundred of the most familiar magazines—such as *Atlantic, Nation, Scientific American, Time.* Probably there won't be much for you in these magazines unless your topic is something like "Van Gogh's Reputation Today," or "Popular Attitudes Toward Surrealism, 1930–40," or "Jackson Pollock as a Counter-Culture Hero of Fifties America," or some such thing that necessarily draws on relatively popular writing. *Readers' Guide* is also available (going back to 1984) on CD-ROM and on the Web.

Art Index, BHA, RAA, RILA, and *ARTbibliographies MODERN* are indexes to many scholarly periodicals—for example, periodicals published by learned societies—and to bulletins issued by museums.

Art Index lists material in about three hundred periodicals, bulletins, and yearbooks; it does not list books but it does list, at the rear, book reviews under the name of the author of the book. *Art Index* covers not only painting, drawing, sculpture, and architecture, but also photography, decorative arts, city planning, and interior design—in Africa and Asia as well as Western cultures. A computerized version of *Art Index* (going back to October 1984) is available through WILSONLINE, and, on CD-ROM and on the Web. The Web version has a unified index for volumes from 1985, a great help.

BHA: Bibliography of the History of Art represents a merger of the next two bibliographies that are discussed, *RAA* and *RILA,* but you will still need to consult these two for material published before 1989. *BHA* covers visual arts in all media *but* it is limited to Western art from

Late Antiquity to the present. Thus, it excludes not only ancient Western art, but also Asian, Indian, Islamic, African, and Oceanic art, and American art before the arrival of Europeans. In addition to including citations of books, periodical articles, exhibition catalogs, and doctoral dissertations, it includes abstracts (*not* evaluations) in English or French. Available on CD-ROM, 1973 to the present, and on the Web.

RAA: *Répertoire d'art* lists books as well as articles in some 1,750 periodicals (many of them European), but, like its successor *BHA*, it excludes non-Western art, and beginning with 1965, it excludes art before the early Christian period. It also excludes artists born after 1920. Included in *BHA* on CD-ROM and on the Web.

RILA: *International Repertory of the Literature of Art* lists books as well as articles in some three hundred journals. It is especially useful because it includes abstracts, but it covers only post-Classical European art (which began with the fourth century) and post-Columbian American art. It thus excludes (again, like its successor *BHA*) prehistoric, ancient, Asian, Indian, Islamic, African, Oceanic, and Native American art. Several volumes of *RILA* can be searched at one time by using its *Cumulative Indexes.* It is available on DIALOG.

ARTbibliographies *MODERN* used to cover art from 1800, but beginning in 1989 it limited its coverage to art from 1900. It now provides abstracts or brief annotations not only of periodical articles concerned with art of the twentieth century, but also of exhibition catalogs and books. Entries since 1974 can be searched through DIALOG. It is also available on CD-ROM.

Architectural Periodicals *Index* indexes about five hundred journals. It is available online as *The Architecture Database* (DIALOG), with files since 1978.

Avery *Index to Architectural Periodicals* is available on CD-ROM and RLG Eureka on the Web.

Major newspapers also cover art topics—especially books and exhibitions. Some major newspapers have their own indexes, for instance **The New York Times *Index*** (1951–) and **Times of London *Index*** (1790–). *The New York Times Index* is available as part of the *Expanded Academic Index.* Many other indexes to newspapers are available on **LEXIS/ NEXIS.** LEXIS/NEXIS is a database containing the full texts, but not the graphics or photos, of thousands of newspapers and magazines (e.g., *Art in America*) from around the world. It is now available to authorized users on the Web as Lexis/Nexis Academic Universe.

Users of the Internet can access (for free) art-historical information from bibliographic and research databases of the Getty Art History Information Program (AHIP), thereby simultaneously searching *Avery Index* (1977–94), *RILA* (1975–89), and the *Provenance Index* (which gives details of ownership from sales catalogs).

Whichever indexes you use, begin with the most recent years and work your way back. If you collect titles of materials published in the last five years, you will probably have as much as you can read. These articles will probably incorporate the significant points of earlier writings. But of course it depends on the topic; you may have to—and want to—go back fifty or more years before you find a useful body of material.

Caution: Indexes drastically abbreviate the titles of periodicals. Before you put the indexes back on the shelf, be sure to check the key to the abbreviations so that you know the full titles of the periodicals you are looking for.

Other Guides

There are a great many reference books, not only general dictionaries and encyclopedias, but also **dictionaries of technical words** (e.g., *Adeline's Art Dictionary*, reissued as *Adeline Art Dictionary* in 1966), **dictionaries of symbolism** (James Hall's *Dictionary of Subjects and Symbols in Art*, second edition, 1979), and **encyclopedias of special fields** (*Encyclopedia of World Art*). *Adeline's Art Dictionary* defines terms used in painting, sculpture, architecture, etching, and engraving. Hall's *Dictionary* is devoted chiefly to classical and biblical themes in Western art: If you look up the Last Supper, you will find two detailed pages on the topic, explaining its significance and the various ways in which it has been depicted since the sixth century. You will learn that in the earliest depictions Christ is at one end of the table, but later Christ is at the center. A dog may sit at the feet of Judas, or Judas may sit alone on one side of the table; if the disciples have haloes, Judas's halo may be black. Again, if in Hall's *Dictionary* you look up *cube,* you will find that in art it is "a symbol of stability on which Faith personified rests her foot; . . . Its shape contrasts with the unstable globe of Fortune."

Two other guides to symbolism—one of them by James Hall—include non-Western material, so their coverage is considerably broader than in Hall's *Dictionary of Subjects and Symbols,* but the coverage of Western material is somewhat thinner. The books are Hans Biedermann's *Dictionary of Symbolism* (1992), and James Hall's *Illustrated Dictionary of Symbols in Eastern and Western Art* (1994).

For definitions of **chief terms** and for **brief biographies,** see *The Oxford Dictionary of Art,* new edition (1997), edited by Ian Chilvers,

Harold Osborne, and Dennis Farr. *The Dictionary of Women Artists* (1997), edited by Delia Gaze, is a two-volume work covering all periods. For artists of the twentieth century, the best general reference works are *Contemporary Artists,* fourth edition (1996), and Ian Chilvers, *Dictionary of Twentieth-Century Art* (1998).

There are also subject bibliographies—books that are entirely devoted to listing (sometimes with comment) publications on specific topics. Yvonne M. L. Weisberg and Gabriel P. Weisberg's *Japonisme* (1987) lists and comments on almost seven hundred books, articles, exhibition catalogs, and unpublished dissertations concerning the Japanese influence on Western art from 1854 to 1910. Similarly, Nancy J. Parezo, Ruth M. Perry, and Rebecca S. Allen's *Southwest Native American Arts and Material Culture* (1991) lists more than eight thousand references, including books, journals, exhibition catalogs, and dissertations. Wolfgang Freitag, *Art Books: A Basic Bibliography of Monographs on Artists,* second edition (1997), provides lists of books for more than two thousand international artists.

How do you find such reference books? The best single volume to turn to is Lois Swan Jones, *Art Information and the Internet: How to Find It, How to Use It* (1998). Updates for Jones's book will be posted at *www.orgxpress.com/artupdate*
Among other useful guides to the numerous books on art are these:

> Donald L. Ehresmann, *Fine Arts: A Bibliographic Guide to Basic Reference Works, Histories, and Handbooks,* 3rd ed. (1990).
>
> Etta Arntzen and Robert Rainwater, *Guide to the Literature of Art History* (1980). This covers material only to 1977; a supplementary volume covering material from 1977 to 1997 is in preparation.
>
> Max Marmor and Alex Ross, *Guide to the Literature of Art History,* volume 2 (scheduled for 1999). This includes updated material as well as chapters on such topics as design history, patronage and collecting, aesthetics, criticism, and theory.

What about reference books in other fields? The best guide to reference books in all sorts of fields is

> *Guide to Reference Books,* ed. Robert Balay, 11th ed. (1996).

There are guides to all these guides: *reference librarians.* When you don't know where to turn to find something, turn to the librarian.

Finally, if you are interested in learning about new bibliographic tools, keep an eye on current issues of a journal devoted to such matters: *Art Documentation* (1982–).

ART RESEARCH ON THE INTERNET AND THE WORLD WIDE WEB[o]

Chances are that you already have used your campus network of computers—for electronic mail, to search your library's catalog, or to view course materials (lecture notes, primary sources, study guides) and to review images. There are compelling reasons to extend your research on an artwork to the Internet as well. **The Internet** is the largest of all computer networks; it is used for various kinds of communication and for the exchange of any data that can be stored digitally.[†] Because it includes *gateways* to other networks, both national and international, its reach is extraordinary. Its value for research has evolved during the last decade from the growth in the subset of the Internet called the **World Wide Web** (www), in conjunction with the introduction of **HTML** (hypertext markup language). This computer language uses a set of codes to integrate text and imagery and to embed cross-references (**links**) to other files that may or may not be stored on the same server or maintained by the same individuals or institutions. Since these links are the addresses or **URLs** (uniform resource locators) of still other Web files, one can view them immediately; for this reason, many people think of links as instantly available footnotes. HTML files often are called *pages* (by analogy to a book or a newspaper), and these pages can be individual components of *sites* (multiple files that belong together on a particular server or are the responsibility of one individual, entity, or institution).

---

[o]This discussion of art research on the Internet and the World Wide Web is by Leila W. Kinney, co-editor of the College Art Association's online journal, *CAA.reviews,* and a member of the Department of Architecture at the Massachusetts Institute of Technology. Ms. Kinney wishes to express her thanks to Jerry Saltzer, professor emeritus of electrical engineering and computer science at MIT, for his generous advice on the technical aspects of this section.

[†]For definitions of technical terms, see Mitchell Shnier, *Dictionary of PC Hardware and Data Communications Terms* (O'Reilly and Associates: 1996). Known as "O'Reilly's," this dictionary also is available online by subscription; see *http://www.ora.com/reference/dictionary/* for details.

The Internet and Research

For students of art, the Internet is useful in three ways. As a **research tool,** it offers a mechanism for searching library catalogs, museum collections, electronic publications (including the news), and combined indices of print publications. As the Internet provides a means of displaying images, texts, and archival materials of all kinds, it functions as a **repository** and the computer (or more precisely, the software on it known as a Web **browser**) functions as a **viewing device** for works of art and architecture. But these new opportunities bring new problems. The vast quantity of material available on the Internet and the huge number of listings that can be searched could make it difficult to locate information on a specific topic. And because "anything goes" on the Internet, not all of it can be trusted. Finally, even if we grant that any reproduction is a poor substitute for the original artwork, the distortions introduced by computer monitors (fuzzy contours, exaggerated contrasts, pixelated textures) make the digital reproductions even less accurate. Rigorous search techniques, careful evaluation of sources, and verification of the physical characteristics of an artwork are essential.

Art created explicitly for exhibition on the Web already has a significant presence, and probably "writing about art" soon will encompass Web projects, electronic imaging, robotic sculptures, and "telematic" performances. For the present, however, the task is to navigate the extensive material on art that has been indexed electronically or placed on the Web. Fortunately, there now are guides to the many art sites on the Web (see pages 194–195). However, electronic technology means continuous change. New resources are added daily, and existing ones are modified, moved, or simply disappear. Conventions for presenting and identifying material still are being established. Thus, you need to understand how the Internet works and what kinds of information about art are to be found there, so that you can take full advantage of the medium.

Where to Start

Because the Internet offers a vast range of information, created for many purposes and many audiences, you cannot assume that online research will save time. For most topics, searching databases of published literature will be more efficient than searching the Web. Major research libraries and guides to periodical literature on art and architecture have been cataloged with research in mind; through **OPACs** (online public access catalogues) you can find publications specific to your topic:

International Directory of Art Libraries
http://iberia.vassar.edu/~art/virtual.html
Libraries via WWW
http://sunsite.Berkeley.EDU/Libweb/

"Bundled" Databases

You may not have to search each catalog individually, however, because unified databases can simultaneously search an international range of materials in print. In addition, databases allow library catalogs, periodical literature, and archival holdings to be located in one search. *WorldCat* (a product of OCLC, the Online Computer Library Center (*http://www.oclc.org/*), for example, claims over 36 million records from libraries worldwide. Files maintained by RLIN, the Research Libraries Information Network (*http://www.rlg.org/rlin.html*) incorporate archives, photographs, posters, and films, which otherwise can be difficult to find. The catch is that these resources are widely but not universally available; institutions subscribe to bibliographic services, select the specific databases they contain, and frequently allow only their own students and faculty to use them. Consult your library's reference desk or Web page to find what resources your institution makes available.

You cannot assume that online databases are exactly the same as the card, print, or even CD-ROM versions to which you may be accustomed. That basic guide to periodical literature, *Art Index,* for example, is now frequently bundled into *FirstSearch* (also from OCLC), where its electronic alter ego is *Art Abstracts;* there it is only one of a group of databases that can be searched simultaneously. *Avery Index to Architectural Periodicals* is another confusing case. The free version available on the Web through the Getty Information Institute contains records from 1977 to 1994 (*http://www.getty.edu/index/ avery.html*). For ongoing coverage, updated daily, your library must subscribe to *The Avery Index* offered by RLG, the Research Libraries Group (*http://www.rlg.org/cit-ave.html*). A similar situation pertains for *RILA, the International Repertory of the Literature of Art.* Records from the years 1975 to 1989 are searchable through the Getty site; its successor *BHA,* the *Bibliography of the History of Art,* with twice the coverage, comes from RLG. You are not, in other words, going to find the most recent literature on art and architecture by using the Web unless your library subscribes to *Avery, BHA,* or similar online indices. Always check the description of a database (often found in the hyperlink attached to its name) to determine the scope of coverage. Finally, do not assume that all your research can be conducted online and that you will not need to go to the library; few databases contain records for materials published before the early 1970s.

Knowing the Search Language

Once you have identified electronic resources, you need to learn the protocols of the specific database or search engine. Boolean operators (AND, OR, and NOT) have been the norm, but the desire for natural language and concept-based queries has led to great variation in search language. It is wise, therefore, to click on "help" for instructions on how to search. Another pitfall for electronic searching has been the notorious lack of standardization in references to artists' names. Fortunately, the Getty has provided an invaluable service in *ULAN*, the *Union List of Artists' Names* (*http://www.gii.getty.edu/unionlist.html*). Should your search fail to turn up literature on a specific artist or architect, check *ULAN* for real names as opposed to pseudonyms or alternative spellings.

Limiting and Expanding Your Search

Most likely, your search of electronic databases will turn up an avalanche of references. For many writing assignments, finding relevant sources will be more important than finding *every* reference to a topic, and the wealth of information can prove more frustrating than liberating. The large number of possible matches to your subject may cause you to curtail your search too soon—"Surely I already have plenty," you think—thus missing what you really needed. In such a case, you must narrow the search by finding more precise subject or keywords. Johannes van der Wolk's article "Vincent van Gogh and Japan" may not emerge from a search for its author's name if individual articles published in exhibition catalogues and symposium proceedings are not indexed. Nor would it appear if you enter a keyword or subject search using "Van Gogh AND Japan" instead of "Van Gogh AND Jap." Why doesn't it appear? Because the subject word "Japonisme," but not "Japan," is used for the volume in which it is published, *Japonisme in Art: An International Symposium* (1980). Indeed, persistence is required to track down some of the most valuable research on art. It is even possible that searching the same database at different times can produce different results: The search engines may not have the entire database available every day; its results may depend on what others have been searching recently; or some of its disks may respond more quickly than others. Yet electronic searches offer many benefits. Their coverage is unparalleled; most of them allow you to e-mail yourself selected references; some will tell you whether your library owns a copy of the book or journal you seek; and others can connect you to full-text online journals or document delivery services.

✔ Checklist for Successful Electronic Searching

In conducting an electronic search for print publications, keep these steps in mind:

✔ Consult your library's Web page or reference desk to learn which electronic databases are available.

✔ Familiarize yourself with each database: Which years does it cover? Which search terms does it use? Which periodicals, library catalogs, or archival holdings does it index?

✔ When you know exactly what you are looking for (a proper name and a specific journal, for example), use the "advanced" search option.

✔ Use the "simple" screen and the subheadings generated by a simple search to specify the topic more precisely in a subsequent advanced search; for example, a simple search can produce the subject entry "Performance Art—United States—With Robots By Women," which then can be followed as a link or reentered as a subject word in an advanced search.

✔ Search the same topic on more than one occasion and in different databases. In each search, use related but varying or truncated subject and keywords (i.e., "Jap" not "Japan" or "*Japonisme*").

✔ Search early so that you can request material through Interlibrary Loan, if available.

✔ E-mail records to yourself and print them, or keep precise notes, in case you need to retrieve items again. Include the references found, the database and search terms used to locate them, and record identification numbers.

✔ Don't rely upon the Internet or computerized databases alone. Few databases contain literature published before 1970, for which you need your library's card catalog and printed indices to periodical literature.

Topics for Which the Web Excels

For some topics, a combination of print sources and information found on Web sites works best. If you are researching an artist as popular as van Gogh, unless you are rigorous and disciplined in your approach, the Internet can be like quicksand. Yet for contemporary art, recent exhibitions, current controversies—indeed, for any topic for which coverage has appeared in the daily press or might appear on a museum's Web site—Internet search engines are advisable. Major U.S. newspapers and many international ones have a Web edition, and many maintain online archives of back issues. Thousands of museums have Web sites. Consider a topic involving the fate of stolen art—specifically, art confiscated by the Nazis during World War II and the controversy about the circumstances of the acquisition of some artworks and antiquities now in prominent collections. The issues are intertwined with stories about the behavior of Swiss banks, Allied troops, and government agencies during and after the war. Questions about the ethical responsibilities of art dealers and museum boards have been raised. For this topic, choosing one or more of the full-text Internet search engines makes sense. They will generate entries that give the first sentence or two of the sites found and will produce many leads, from registers of lost works to reviews of the latest books, from press releases by museums or governments to college courses on the history and cultural policies of the Third Reich. Casting such a wide net can be effective.

The Web as a Repository

Although the Internet encourages continuous change, its vast storage capacity and relatively unencumbered means of distribution also encourage its use as a permanent repository for cultural material. Web sites devoted to early civilizations (Greek, Roman, medieval, Byzantine) are particularly impressive, and large collections of historical documents and literature no longer in copyright can be found. In artistic terms the Web is considered an unconventional and experimental space. Sites devoted to art movements such as surrealism and performance art were among the first to be established. Given the difficulties in documenting (not to mention viewing) performance art, the archived and netcast performances available at the Franklin Furnace Web site are an exceptional resource (*http://www.franklinfurnace.org/*). And some museums commission multimedia art or feature contemporary artists on their Web sites in conjunction with exhibitions.

Internet Search Engines

Searching the Web requires even more know-how than searching databases of print literature about art. In particular, it is important to understand the implications of machine-driven (robotic) Internet search engines as opposed to those that rely upon people to categorize Web sites by subject. In general, the former will alert you to anything and everything (including multiple references to the same site) and the latter will be more likely to lead you to pertinent sites. AltaVista (*http://www .altavista.com/*) is an example of the former and *Yahoo* (*http://www .yahoo.com/*) of the latter. The librarians at the University of California, Berkeley, have produced an excellent set of tutorials and tables that explain how to analyze your topic and decide which search engines are most appropriate for your needs (*http://lib.berkeley.edu/TeachingLib/ Guides/Internet/FinInfo.html*).

Online Directories

As the Internet becomes more commercialized and compartmentalized, you may prefer to use annotated directories tailored to scholarly and educational resources. These directories are lists of links selected and ordered by subject, and often they are compiled by scholars. Yet even these directories can be mind-boggling as they may contain a mixed bag of Web projects and addresses that seem to be included only because they were submitted and deemed to belong to the category "art." Nonetheless, these directories are essential for locating materials unique to the digital realm, including the many educational, archival, and experimental projects on art that are created expressly for Web dissemination.

Directories of Art-Related Web Sites

ADAM: Art, Design, Architecture & Media Information Gateway
http://adam.ac.uk

History of Art Department, Birkbeck College, London, *The History of Art Virtual Library*
http://www.hart.bbk.ac.uk/VirtualLibrary.html

Lois Swan Jones, *Art Information and the Internet: How to Find It, How to Use It* (1999), updates at
http://www.oryxpress.com/artupdate

Mary Molinaro, *ArtSource*
http://www.uky.edu/Artsource/artsourcehome.html

Christopher L.C.E. Witcombe, *Art History Resources on the Web*
http://witcombe.bcpw.sbc.edu/ARTHLinks.html

Directories of Museums

Art Museum Network
http://www.amn.org/

Elsas Producties (Links to Museum Sites)
http://www.elsas.demon.nl/

Musée
http://www.musee-online.org/

The World Wide Web Virtual Library: Museums Pages
http://www.icom.org/vlmp/

Web Pages for Art History Courses

Robert Derome, *Index of Art History Courses on the Web*
http://www.er.uqam.ca/nobel/r14310/Cours/cours.html

Posting Questions

When you have a topic so specialized or novel that the subject is not likely to be indexed, your instructor, reference librarian, or slide curator may want to post a query to one of the electronic discussion groups devoted to topics in the visual arts that operate by e-mail (known as *listservs*). It is not good form, however, to ask for help with questions that can be answered by intelligent use of library resources, or to join a group just to ask one question. Archives of topics previously discussed (called *threads*) are covered by full-text Internet search engines, and you should examine relevant messages before repeating a question already on the list.

Finding and Viewing Images

In addition to using the Internet for research, students of art history use it to view reproductions. In fact, the ability to digitize images has been a catalyst for the development of the Web, and the notion of virtual museums has inspired a genre of display. Image files are recognizable by their suffixes (usually *.jpg, .jpeg,* or *.JPG,* for Joint Photographic Experts

Group, and .*gif* or .*GIF* for Graphics Interchange Format or Graphics Image File). Yet the technology for searching and indexing images on the Web is just now being developed; see *Arthur*, ART Media and Text HUb and Retrieval System (*http://www.gii.getty.edu/index/databases.html*). If you know the museum or collection where the work resides, it is relatively simple to check its Web site for a reproduction. However, with the notable exception of the Fine Arts Museums of San Francisco (*http://www.thinker.org/*), most museums display online only a small sample of their holdings. If a reproduction cannot be found in a book or a slide collection, the best recourse may be the specialized directories mentioned above; they also will guide you to reproductions of architecture, which are dispersed throughout the Web.

Digital reproductions of artworks have been placed on the Internet for many different purposes. Because some sites that contain art are unedited, idiosyncratic, or partial, you must be alert to several factors when consulting them.

The quality and accuracy of digital reproductions vary enormously. Consider how most digital reproductions of art are made. Many are second or third generation copies; a scan taken from a 35mm slide taken from a print in a book introduces all sorts of degradation. The scanning software allows endless possibilities for adjustment; alterations of color, tonality, and sharpness may improve an image but diminish its resemblance to the original. Digital images—or the computer monitors on which we view them—typically blur contours and surface qualities; tonal contrasts are greater, and relative scale is difficult to discern. Even when high quality scans are available, low-resolution "thumbnails" often are the only reproduction placed on the Web (in order to respect copyright laws). The bottom line: Always try to view a work of art in person; when that is not possible, compare reproductions (those in books, from slides, and on the Internet), keeping in mind their limitations. In spite of these drawbacks, however, viewing previously unphotographed art, bodies of related works newly brought together, or multiple views of sculpture and architecture—all available on the Web—can add substantially to your knowledge.

The authorship of the site can affect the reliability of its information. National museums and libraries generally have exemplary Web presentations, because they already have well-documented, comprehensive collections and a public mission to make their holdings accessible. Many private museums in this country and abroad document the images they display accordingly. In addition, educational institutions have seen the Web as an opportunity to build digital image archives for teaching and research. Not that institutional authorship is always preferable; some museums limit their Web sites to publicity materials for temporary exhibitions, while many of the best art sites are created by individual scholars in the field.

✔ Checklist for Evaluating Web Sites

The anonymity of many Web sites, the tendency to substitute corporate for individual identity, and the preeminence given to the address rather than to the creators of Web files all complicate the task of evaluating the site you have located and the source of its authority for your topic. Consider these questions in determining an art site's value for your paper:

- ✔ Who produced the site (a museum, a teacher, a commercial entity, a student)?
- ✔ Is it authoritative enough to cite? Sites consisting of review materials for an art history course probably will not offer information of sufficient depth for writing a paper, nor will an online tour of a museum collection. They are the equivalent of a class handout or a museum label.
- ✔ Are the sources of information indicated and verifiable? If they are, are they specialized? (Some sites reprint passages from online encyclopedias or standard reference books as their explanatory text. When is the last time you got an "A" for quoting an encyclopedia?)
- ✔ Is the information you have located unique? If the site is based on research papers or scholarly books, go to those sources, which will offer more substance and detail.
- ✔ Are the reproductions accurate and documented? Compare them to a slide or a print from a book.
- ✔ What is the likely permanence of the materials you are viewing? Are they promotional or part of a course assignment—something that might be gone by the time your reader tries to retrieve it?
- ✔ Does the purpose of the site fit the reason you are citing it? Some sites are created for long-term reference; others present current information or ephemeral events. Citing online materials for an architectural competition makes sense if you are writing about that competition, but a museum's current exhibitions page is probably not a reliable source for an image of an artist's work.
- ✔ Can the information contained there be confirmed by other sources?

Referencing Web Pages

Once you have satisfied yourself about the quality and appropriateness of your sources, how should you cite them? Reference formats for print publications have been adopted and modified to cite electronic sources (see the examples below). The problem is that Web sites themselves are not documented accordingly. Unfortunately, there is no standardized title page for Web sites. Nor is there a header or footer format that provides all of the necessary information (even conventions for URLs are changing). Nor are browsers configured to record precisely the information you will need to write a proper citation. And the very nature of hypertext, which allows you to move from site to site, not just from the beginning to the end of the one with which you began, can make it easy to lose track of the path you followed. Keeping in mind certain procedures when you begin your research may save countless hours of trying to retrace your steps when you want to prepare footnotes or bibliography.

Bookmarking. Use the command in your browser that records the title and URL of the Web page you are viewing. Print the page for reference, if possible. However, current browsers do not record information adequately for proper citation. Citing the URL is not sufficient.

✔ Checklist for Electronic Documentation

The documentation you need for a reference probably will have to be gathered from various places on the site. Before leaving a site, make sure you have recorded these facts:

✔ Author and institution hosting the site. When no author is indicated, record the webmaster or e-mail address to be contacted about the page.

✔ Title of the page(s) you are referencing and URL. If a title is not provided, record the file name (everything behind the last forward slash in the URL).

✔ For digital images, the identifying information: artist, title, date, owner or location (as in "Kyoto National Museum, Japan") and URL.

✔ Name of the site to which the pages you want to cite belong, especially if it is an extensive one.

✔ Revision date for the page or site, if stated, and the date you viewed it.

✔   Before leaving the site, ask yourself: If the site moves or reorganizes its Web pages (in other words, if the URL changes), have I provided enough information for a reader to find this page with the help of a search engine?

Not all of this information may be available or necessary for every citation you make, but it will save time in the end to record everything you can find at the outset.

Special Characteristics of Frames, Mirrors, and Sites Powered by Search Engines.   When you are in a *frames* Web site (one with multiple windows), be sure to learn the URLs of any linked sites that you call up within a frame. (Your browser may retain the address of the host site and not give the one to which you have just linked; techniques will vary with the browser.) Sites powered by search engines and those duplicated or *mirrored* at other locations also present difficulties. The addresses of underlying pages will vary depending upon the search results or whether they are accessed at mirror sites. For such unstable URLs, it is preferable to offer the top-level directory or original site address. The following reference formats have been devised with these factors in mind.

Citations for Electronic Materials

A Footnote or Endnote Reference for a Site

> 1. Nigel Strudwick, Egyptology Resources, The Isaac Newton Institute for Mathematical Sciences, Cambridge University, http://www.newton.cam.ac.uk/egypt/, 7 July 1998.

A reference should begin with the author's name, followed by the title of the site, underlined or in italics. The name of the institution hosting the site (equivalent to the publisher of a book) follows, then comes the URL, which consists of the protocol "http" and address (everything following the double slashes), underlined or in italics. Finally, record the date on which you consulted the site. If the host's identity is not displayed on the front or home page of the site, usually it can be deciphered from the domain name. Entering the first part of the address—up to the first single slash—will take you to the domain in question. In this case "newton" stands for the Isaac Newton Institute for Mathematical Sciences, "cam" for Cambridge University, "ac" for academic institution (the type of domain), and "uk" for United Kingdom (the country). The date of publication frequently will not be indicated, but it should be included when it is.

A Footnote or Endnote Reference for an Online Exhibition

2. Ecole nationale supérieure des beaux-arts, Paris, The Prix de Rome Contests in Painting, Ministry of Culture and Communication, France, http://www.culture.fr/ENSBA/Rome.html, 8 July 1998.

Online exhibitions often do not identify the curators or contributors, as do printed exhibition catalogs; in such cases the name of the museum or institution sponsoring the event will suffice. When the dates of the exhibition are provided, they should be listed after the title.

A Footnote or Endnote Reference for a Digital Image

3. Jan Van Eyck, The Arnolfini Portrait, 1434, National Gallery, London,   http://www.nationalgallery.org.uk/collection/sainsbury/img/ E0186.JPEG, in The National Gallery: The Collection: The Sainsbury Wing, http://www.nationalgallery.org.uk/collection/sainsbury/index.html, 15 January 1999.

For a digital reproduction, the artist's name is followed by the title. Then comes the URL of the image, the date ("n.d." when the work is undated), and collection. Note that the address of the page on which the image is found also is included, because it is likely that a search could find the web page but not the image file linked to it.

For bibliographies, changes are necessary in the order of the author's name (last name first) and the punctuation. In bibliographic format, the above entries should be listed as follows:

Reference to a Site in a Bibliography

Strudwick, Nigel. Egyptology Resources. The Isaac Newton Institute for Mathematical Sciences, Cambridge University. http://www .newton.com.ac.uk/egypt/. 7 July 1998.

Reference to an Online Exhibition in a Bibliography

> Ecole nationale supérieure des beaux-arts, Paris. The Prix de Rome Contests in Painting. Ministry of Culture and Communication, France. http://www.culture.fr/ENSBA/Rome.html. 8 July 1998.

Reference to a Digital Image in a Bibliography

> Eyck, Jan van. The Arnolfini Portrait, 1434. National Gallery, London. http://www.nationalgallery.org.uk/collection/sainsbury/img/E0186 JPEG. In The National Gallery: The Collection: The Sainsbury Wing.   http://www.nationalgallery.org.uk/collection/sainsbury/ index.html. 15 January 1999.

Further instructions can be found in Janice R. Walker and Todd Taylor, *The Columbia Guide to Online Style* (1998), with updates at *http://www.columbia.edu/cu/cup/cgos/.*

Embedding URLs in Your Paper

If you use electronic sources, you may be asked to hand in both a printed copy of your paper and a word processing file with the URLs of the sites you have referenced embedded in the text. Most word processors offer this direct linking from a text file to a Web address, which allows the URLs in your manuscript to become hyperlinks. When you cite sources in this way, use the appropriate command in your word processor to insert the URL between parentheses in your text, and add a footnote or endnote to provide the full documentation outlined above. Your instructor will then be able to follow the links from your file without having to retype the addresses. Your file can be submitted either on disk or as an attachment to e-mail, but it would be courteous to ask your instructor whether he or she would appreciate this convenience. In some cases writing a paper on art may involve creating a Web page with hyperlinks to digital images and electronic sources; that, however, is a topic for another book.

READING AND TAKING NOTES

As you read, you will find references to other publications, and you will jot these down so that you can look at them later. It may turn out, for example, that a major article was published twenty years ago and that the most recent writing about your topic is a series of footnotes to this piece. You will have to look at it, even though common sense had initially suggested (incorrectly, it seems) that the article would be out of date.

A few words about reading sources:

- Before assiduously taking notes from the first paragraph onward, or highlighting long passages, it is usually a good idea to scan a work in order to get a sense of the thesis. You may, for instance, find an early paragraph that states the thesis, and you probably will find a concluding paragraph that offers a summary of the evidence that supports the thesis. Having gained from this preview a general idea of the content of the work, you can now highlight sparingly while you read the material carefully and critically.

- Reading critically does not mean reading hostilely, but it does mean reading thoughtfully, continually asking yourself whether the assertions are adequately supported with evidence. Be especially sure to ask what can be said against assertions that coincide with your own beliefs. One can almost say that the heart of critical thinking is a willingness to face objections to one's own beliefs.

- When the book or journal you are reading is your own, the best way to read critically—interactively rather than passively—is to annotate it, jotting down brief responses ("but see . . . ," "?" and so forth) in the margins. If you highlight, try to confine yourself to brief passages; there is no point in highlighting whole pages. If you don't own the source, don't annotate it and don't highlight it. Either photocopy it, if it is fairly brief, and then mark up your copy—consider reducing the copy so that you will have ample margins for your comments—or take notes in a notebook, on a word processor, or on 4 × 6-inch cards or slips of paper. (Although it surely is best to write the paper on a word processor, many writers prefer to take their notes on cards.)

Here are guidelines for note-taking:

1. **When you take notes on 4 × 6 cards, write on one side of the card only;** material on the back of a card is usually neglected when you come to write the paper. If your notes from a source on a particular point run to more than one card (say, to three), number each card: 1 of 3, 2 of 3, 3 of 3. Use 4 × 6 cards because the smaller cards (3 × 5) are too small

for summaries of useful material, and the larger cards (5 × 7) invite you to put too much material on one card.

2. **Write summaries rather than paraphrases;** write abridgments rather than restatements, because restatements may turn out to be as long as or longer than the original. There is rarely any point to paraphrasing; generally speaking, either quote exactly (and put the passage in quotation marks, with a notation of the source, including the page number or numbers) or summarize, reducing a page or even an entire article or chapter of a book to a single 4 × 6 card or to a few sentences. Even when you summarize, indicate your source (including the page numbers) on the card, so that you can give appropriate credit in your paper.

3. In your summary you will sometimes quote a phrase or a sentence—putting it in quotation marks—but **quote sparingly.** You are not doing stenography; rather, you are assimilating knowledge and you are thinking, so for the most part your source should be digested rather than engorged whole. Thinking now, while taking notes, will also help you later to avoid plagiarism. If, on the other hand, when you take notes you mindlessly copy material at length, later when you are writing the paper you may be tempted to copy it yet again, perhaps without giving credit. Similarly, if you photocopy pages from articles or books and then merely underline some passages, you probably will not be thinking; you will just be underlining. But if you make a terse summary on the word processor or on a note card, you will be forced to think and to find your own words for the idea.

Most of the direct quotations that you copy should be effectively stated passages or crucial passages or both. In your finished paper these quotations will provide authority and emphasis. Be sure not to let your paper degenerate into a string of quotations.

4. If you quote but omit some material within the quotation, be sure to indicate the omission by an ellipsis, or three spaced periods (as explained on pages 236–37). **Check the quotation for accuracy,** and check the page number you have recorded on your card.

5. **Never copy a passage by changing an occasional word,** under the impression that you are thereby putting it into your own words. Notes of this sort may find their way into your paper, your reader will sense a style other than your own, and suspicions of plagiarism may follow. It is worth saying yet again that you should either quote exactly, and enclose the words within quotation marks, or summarize drastically. In both cases, be sure to give credit to your source. (For a detailed discussion of plagiarism, see pages 238–42.)

> *van G's* <u>Portrait of Tanguy</u>
> (1887-1888)
>
> Picture includes copies of Japanese prints but <u>not</u> in style of prints: lacks characteristic contour lines of J. prints
> Why?
>   Pollock and Orton claim V's interest is in prints' <u>subject-matter</u>, not in their style: "Attention is drawn to what they depict and not to the stylistic character of the woodcuts" (p. 41)
>
>   ((seems to me van G is concerned with their style))

    6. **Jot down your own responses to the note.** For example, you may want to say, "Baker made same point 5 yr earlier." But make certain that later you will be able to distinguish between these comments and the notes summarizing or quoting your source. A suggestion: Surround all comments recording your responses with double parentheses, thus: ((. . .)).

    7. In the upper corner of each note card, **write a brief key**—for example, "van G's Portrait of Tanguy"—so that later you can tell at a glance what is on the card. See the sample note card above. Things to notice are: a brief heading at the top identifies the subject of the card; the quotation is followed by the page citation; the writer's own response to the quotation is enclosed within double parentheses so that it cannot be confused with the writer's notes summarizing the source.

    8. **Record the title of your source accurately.** Index cards were once a great way to record sources, and many art historians continue to use them, but now, with word-processing and database software, you can more easily keep track of sources. When using the word processor, keep all of your references in one file and use a standard bibliographic format such as this one:

Harper, Prudence Oliver. <u>The Royal Hunter: Art of the Sassanian Empire</u>. New York: Asia Society, 1978.

The computer will allow you to sort the list, at least by the first word (author's last name) in each entry. As you write your paper, you can insert ("copy and paste") each reference into the appropriate footnote or endnote (making the appropriate changes in format, as needed).

If you envision using many sources, a database—which provides much more power than word-processing software—may offer a better approach. Many programs come with a template for references so that all you have to do is enter the information in the preexisting fields. Then you can sort or limit the references by title, author, source, or by any of the ideas you have drawn from the source, such as the work of art the source refers to, the artist it discusses, or the dates in history that it covers. For example, when writing a long paper on the sculptures from the Parthenon (the so-called Elgin Marbles), a reference in your database might have the following usable fields:

Author:	John Henry Merryman	Period:	Classical
Title:	Thinking about the Elgin Marbles	Dates:	440-432 BC
Source:	Michigan Law Review	Artist:	Phidias
Cite:	Vol. 83, 1985, pp. 1881-1923	Topic 1:	Classical sculpture
Art:	Greek	Topic 2:	Cultural repatriation

Quote 1:   "Elgin removed (or took from the ground where they had fallen or from the fortifications and other structures in which they had been used as building materials) portions of the frieze, metopes, and pediments" (pp. 1883-1884).

Quote 2:   "Despite national laws limiting the export of cultural property, some of it still finds its way abroad" (p. 1889).

Quote 3:   "That, in the end, is what the law and politics of cultural property are about: the cultural heritage of all mankind" (p. 1923).

DRAFTING AND REVISING THE PAPER

There remains the difficult job of writing up your findings, usually in two thousand to three thousand words, or eight to twelve double-spaced typed pages. Here is some advice.

1. **Reread your notes,** sorting cards into packets by topic, or moving blocks if your notes are on a word processor. Put together what belongs together. If your notes are on a word processor, print them out, scissor them apart, and then arrange them into appropriate groups and sequences. Don't hesitate to reject material that—however interest-

ing—now seems redundant or irrelevant. In doing your research you quite properly took many notes (as William Blake said, "You never know what is enough unless you know what is more than enough"), but now, in looking your material over, you see that some of it is unnecessary and you reject it. Your finished paper should not sandbag the reader; keep in mind the Yiddish proverb, "Where there is too much, something is missing."

After sorting and resorting, you will have a kind of first draft without writing a draft. This business of settling on the structure of your work—the composition of your work, one might say—is often frustrating. Where to begin? Should this go before that? But remember that great artists have gone through greater struggles. If we look at a leaf from one of Raphael's sketchbooks, we find him trying, and trying again, to work out the composition of, say, the Virgin and Child. Nicolas Poussin used a miniature stage, on which he arranged and rearranged the figures in his composition. An X-ray of Rembrandt's *The Syndics of the Cloth Drapers Guild* reveals that he placed the servant in no fewer than four positions (twice at the extreme right, once between the two syndics at the right, and finally in the center) before he found a satisfactory arrangement. You can hardly expect, at the outset, to have a better grasp of your material than Raphael, Poussin, and Rembrandt had of theirs. What Degas said of a picture is true of a research paper: "A good picture requires as much planning as a crime."

2. **From your packets of cards or your rearranged notes you can make a first outline.** In arranging the packets into a sequence, and then in sketching an outline (see pages 92–93), you will be guided by your *thesis*, your point. Without a thesis you will have only a lot of notes, not an essay. This outline or map will indicate not only the major parts of the essay but also the subdivisions within these parts. Do not confuse this type of outline with a paragraph outline (an outline made by jotting down the topic idea of each paragraph); when you come to write your essay, a single heading in your outline may require two or three or more paragraphs.

Don't scorn the commonest organization:

- introduction of the works to be studied, and of the thesis (the thesis is your central idea, your main point)
- presentation of evidence, with interpretation relating it to the thesis
- presentation of counterevidence, and rebuttal
- conclusion

You may find that this organization doesn't suit your topic or you. Fine, but remember that your reader will need to be guided by some sort of organization that you will have to adopt and make clear.

3. **When you write your first draft, leave lots of space at the top and bottom of each page** so that you can add material, which will be circled and connected by arrows to the proper place. On reading your draft you may find that a quotation near the bottom of page 6 will be more appropriate if it is near the top of page 6. Circle it, and with an arrow indicate where it should go. If it should go on page 4, scissor it out and paste it on page 4. This process is a bit messy, but you will get a strong sense of what your paper sounds like only if you can read a draft with all of the material in the proper place. When you relocate material in your draft, also move your note cards containing the material, so that if for some reason you later have to doublecheck your notes, you can find your source easily. Of course, if you are using a word processor, you can move passages of text without literally cutting and pasting.

Your opening paragraph—in which you usually will define the problem and indicate your approach—may well be the last thing that you write, for you may not be able to enunciate these ideas clearly until you have learned something from drafting the rest of your essay. (On opening paragraphs, see pages 142–44.)

4. **Write or type your quotations, even in the first draft, exactly as you want them to appear in the final version.** If you took notes on a word processor, just move the quotations from your notes into the paper.

Short quotations (fewer than five lines of prose) are enclosed within quotation marks but are not otherwise set off.

Long quotations are treated differently. Some instructors recommend that you triple space before and after the quotation and single space the quotation. Other instructors recommend that you double space before and after the quotation, indent the entire quotation ten spaces, and double space it. In either case, do not enclose within quotation marks a quotation that is set off as a block.

(For more on quotations, see pages 235–37.)

5. **Include, right in the body of the draft, all of the relevant citations** (later these will become footnotes or endnotes) so that when you come to revise, you won't have to start hunting through your notes to find who said what and where. If you are using a word-processing program, it will probably allow you to write the footnotes immediately after the quotations, and later it will print them at the bottom of the appropriate

pages. Further, most word-processing programs let you add and delete notes while the computer automatically renumbers the notes in the final version. If, however, you are writing by hand, or on a typewriter, enclose these citations within diagonal lines, or within double parentheses—anything at all to remind you that they will be your footnotes.

6. **Identify works of art as precisely as possible.** Not "Rembrandt's *Self-Portrait*" (he did more than sixty), or even "Rembrandt's *Self-Portrait* in the Kunsthistorisches Museum, Vienna" (they own at least two), but "Rembrandt's *Self-Portrait of 1655*, in the Kunsthistorisches Museum, Vienna," or, more usually, "Rembrandt's *Self-Portrait* (1655, Kunsthistorisches Museum, Vienna)." If the exact date is unknown, preface the approximate date with *ca.*, the abbreviation for *circa*, Latin for "about." Example: ca. 1700. Be sure to identify all illustrations with a caption, giving, if possible, artist, title, date, medium, size, and present location. (See pages 222–23 for more on captions and illustrations.)

7. **Beware of the compulsion to include all of your notes in your essay.** You have taken all these notes, and there is a strong temptation to use them all. But, truth to tell, you now see that many are useless. Conversely, you will probably find as you write your draft that here and there you need to check a quotation or to collect additional examples. Probably it is best to continue writing your draft, if possible; but remember to insert the necessary material after you get it.

8. **Do not let your paper become a string of quotations.** As you revise your draft, make sure that you do not merely tell the reader "X says . . . , Y says . . . , Z says . . . ." When you write a research paper

- you are not merely setting the table with other people's dinnerware; you are cooking the meal
- you must have a point, an opinion, a thesis, a controlling idea

You are working toward a conclusion, and your readers should always feel they are moving toward that conclusion (by means of your thoughtful evaluation of the evidence) rather than reading an anthology of commentary on the topic. Thus, because you have a focus, you should indicate the direction of your thinking by using such formulas as:

> There are three common views on. . . . The first two are represented by A and B; the third, and by far the most reasonable, is C's view that . . .
>
> A argues . . . but . . .
>
> Although the third view, C's, is not conclusive, still . . .

Moreover, C's point can be strengthened when we consider a piece of evidence that she does not make use of . . .

9. **When you introduce a quotation, let the reader see the use to which you are putting it.** "*A* says" is of little help; giving the quotation and then following it with "thus says *A*" is even worse. *You need a lead-in* such as "*A* concisely states the common view," "*B* calls attention to a fatal weakness," "Without offering any proof, *C* claims that," "*D* admits," "*E* rejects the idea that . . . ." In short, it is usually advisable to *let the reader know why you are quoting* or, to put it a little differently, how the quotation fits into your argument. Equally important: After giving a quotation, you'll almost surely want to develop (or take issue with) the point made in the quotation.

It is usually desirable in your lead-in to name the author of the quotation, rather than to say something like "One scholar has said . . ." or "Another critic claims that . . . ." In all probability the authors whom you are quoting are known in the field and your reader should not have to turn to the footnotes to find out whose words he or she has been reading.

Remember, too, that a summary of a writer's views may actually be preferable—briefer and more lively than a long quotation. (You'll give credit to your source, of course, even though all of the words in the summary are your own.) You may have noticed while you were doing your research that the most interesting writers persuade the reader of the validity of their opinions in several ways:

- by letting the reader see that the writer knows what of significance has been written on the topic
- by letting the reader hear the best representatives of the major current opinions, whom they correct modify, or confirm
- by advancing their opinions, and by offering generalizations supported by concrete details

Adopt these techniques in your own writing.

✍ A RULE FOR WRITERS:

Your job is not to report what everyone says but to establish the truth or at least the strong probability of a thesis.

Your overall argument, then, is fleshed out with careful summaries and with effective quotations and with judicious analyses of your own, so that by the end of the paper your readers not only have read a neatly typed paper, but they also are persuaded that under your guidance they have seen the evidence, heard the arguments justly summarized, and reached a sound, thoughtful conclusion. They may not become better persons, but they are better informed.

Consider preparing a dual outline of the sort discussed and illustrated on page 93, in which you indicate what each paragraph *says* and what each paragraph *does*. Thus, typical headings of the second sort will include "Introduces thesis," "Supports generalization," "Uses a quotation as support," "Takes account of an opposing view," and "Summarizes two chief views." Then look over your outline to see if each paragraph is doing worthwhile work, and doing it in the right place.

10. **Make sure that in your final version you state your thesis early,** perhaps even in the title (not "Van Gogh and Japanese Prints" but "Van Gogh's Creative Misunderstanding of Japanese Prints"), but if not in the title, almost certainly in your first paragraph.

11. When you have finished your paper **prepare a final copy that will be easy to read.** Type the paper (see pages 221–22), or print a copy from the word processor, putting the footnotes or endnotes into the forms given on pages 243–53.

A bibliography or list of works cited—see pages 250–53 (*Chicago Manual* style) or pages 260–63 (*Art Bulletin* style)—is usually appended to the research paper, partly to add authority to your paper, partly to give credit to the writers who have helped you, and partly to enable readers to look further into the primary and secondary material if they wish. But if you have done your job well, the reader will be content to leave things where you left them, grateful that you have set things straight.

Or straight for a little while; no one expects more. Writers try to do their best, but they cannot hope to say the last word. As Leo Steinberg says, introducing his comment on Velázquez's *Las Meniñas:*

> Writing about a work such as *Las Meniñas* is not, after all, like queuing up at the A&P. Rather, it is somewhat comparable to the performing of a great musical composition of which there are no definitive renderings. The guaranteed inadequacy of each successive performance challenges the interpreter next in line, helping thereby to keep the work in the repertoire. Alternatively, when a work of art ceases to be discussed, it suffers a gradual blackout. (*October* 19 [1981]: 48)

✔ Checklist for Reviewing a Draft

After you have written a draft and revised it at least once (better, twice or more), reread it with the following questions in mind. If possible, ask a friend to read the draft, along with the questions, or with the peer review checklist on page 123. If your answers or your friend's are unsatisfactory, revise.

✔ Exactly what topic are you examining, and exactly what thesis are you arguing?

✔ Does the paper fulfill all of the promises that it makes or implies?

✔ Are the early paragraphs interesting, and do they give the reader a fairly good idea of what will follow?

✔ Is evidence offered to support arguable assertions?

✔ Has all irrelevant material—however interesting—been deleted?

✔ Are quotations introduced with helpful lead-ins? Are all quotations instructive and are they no longer than they need to be? Might summaries of some long quotations be more effective than the quotations themselves?

✔ Are *all* sources properly cited?

✔ Is the organization clear, reasonable, and effective?

✔ Is the final paragraph merely an unnecessary restatement of what by now, at the end of paper, is obvious? Or is it an effective rounding-off of the paper?

✔ Is the title interesting and informative? Does it create a favorable first impression?

9

Manuscript Form

BASIC MANUSCRIPT FORM

Much of what follows is nothing more than common sense. Unless your instructor specifies something different, you can adopt these principles as a guide to basic manuscript format.

    1. **Use 8½ × 11-inch paper of good weight.** Do not use paper torn out of a spiral notebook; ragged edges distract a reader. If you have written your essay on a computer and have printed it on a continuous roll of paper, remove the strips from the sides of the sheets and separate the sheets before you hand in the essay.

    2. **Write on one side of the page only.** If you typewrite, double-space, typing with a reasonably fresh ribbon. If you submit a handwritten copy, use lined paper and write, in black or dark blue ink, on every other line if the lines are closely spaced. A word to the wise: Instructors strongly prefer typed or printed papers.

    3. **Put your name, instructor's name, class or course number, and the date** in the upper right-hand corner of the first page. It is a good idea to put your name in the upper right corner of each subsequent page, so the instructor can easily reassemble your essay if somehow a page gets detached and mixed with other papers.

    4. **Center the title of your essay** about two inches from the top of the first page. Capitalize the first letter of the first and last words of your title, the first word after a semicolon or colon if you use either one, and the first letter of all the other words except articles (*a, an, the*), conjunctions (*and, but, or,* etc.), and prepositions (*about, in, on, of, with,* etc.), thus:

The Renaissance and Modern Sources of Manet's <u>Olympia</u>

(Some handbooks advise that prepositions of five or more letters—*about, beyond*—be capitalized.) If you use a subtitle, separate it from the title by means of a colon:

Manet's <u>Olympia</u>: Renaissance and Modern Sources

Notice that your title is neither underlined nor enclosed in quotation marks, but when your title includes the title of a work of art (other than architecture), that title is underlined to indicate italics.

5. **Begin the essay an inch or two below the title.** If your instructor prefers a title page, begin the essay on the next page.

6. **Leave an adequate margin**—an inch or an inch and a half—at top, bottom, and sides, so that your instructor can annotate the paper.

7. **Number the pages consecutively,** using arabic numerals in the upper right-hand corner. If you give the title on a separate page, do not number that page; the page that follows it is page 1.

8. **Indent the first word of each paragraph** five spaces from the left margin.

9. **When you refer to a work illustrated in your essay, include helpful identifying details** in parentheses:

> Vermeer's <u>The Studio</u> (Vienna, Kunsthistorisches Museum, Figure 6) is now widely agreed to be an allegorical painting.

But if in the course of the essay itself you have already mentioned where the picture is, do not repeat such information in the parenthetic material.

10. **If possible, insert photocopies of the illustrations, with captions, at the appropriate places** in the paper, unless your instructor has told you to put all of the illustrations at the rear of the paper. Number the illustrations (each illustration is called a *Figure*) and give captions that include, if possible, artist (or, for anonymous works, the culture), title (underlined), date, medium, dimensions (height precedes width), and present location. Here are examples:

> Figure 1. Japanese, <u>Flying Angel</u>, second half of the eleventh century. Wood with traces of gesso and gold, $33\frac{1}{2}'' \times 15''$. Museum of Fine Arts, Boston.

> Figure 2. Diego Velázquez, <u>Venus and Cupid</u>, ca. 1644-48. Oil on canvas, $3'11\frac{7}{8}'' \times 5'9\frac{1}{4}''$. National Gallery, London.

Note that in the second example the abbreviation *ca.* stands for *circa,* Latin for "about."

If there is some uncertainty about whether the artist created the work, precede the artist's name with *Attributed to.* If the artist had an active studio, and there is uncertainty about the degree of the artist's involvement in the work, put *Studio of* before the artist's name.

Four more points about captions for illustrations:

- If you use the abbreviation BC ("before Christ") or AD ("anno domini," i.e., "in the year of our Lord"), do not put a space between the letters.
- BC follows the year (7 BC) but AD precedes the year (AD 10).
- The abbreviations BC and AD are falling out of favor, and are being replaced with BCE ("Before the Common Era") and CE ("Common Era," i.e., common to Judaism and Christianity). Both follow the year.
- Some instructors may ask you to cite also your source for the illustration, thus:

> Figure 3. Japanese, Head of a Monk's Staff, late twelfth century. Bronze, 9¼". Peabody Museum, Salem, Massachusetts. Illustrated in John Rosenfield, Japanese Arts of the Heian Period: 794-1185 (New York: Asia Society, 1967), p. 91.

If, in your source, the pages with plates are unnumbered, give the plate number where you would ordinarily give the page number.

11. **If a reproduction is not available,** be sure when you refer to a work to tell your reader where the work is. If it is in a museum, give the acquisition number, if possible. This information is important for works that are not otherwise immediately recognizable. A reader needs to be told more than that a Japanese tea bowl is in the Freer Gallery. The Freer has hundreds of bowls, and, in the absence of an illustration, only the acquisition number will enable a visitor to locate the bowl you are writing about.

12. **Make a photocopy of your essay, or print a second copy** from the computer, so that if the instructor misplaces the original you need not write the paper again.

13. It's a good idea to **keep notes and drafts,** too, until the instructor returns the original. Such material may prove helpful if you are asked to revise a paper, substantiate a point, or supply a source that you inadvertently omitted.

14. **Fasten the pages of your paper with a staple or paper clip** in the upper left-hand corner. (Find out which sort of fastener your instructor prefers.) Stiff binders are unnecessary; indeed they are a nuisance to the instructor, for they add bulk and they make it awkward to write annotations.

SOME CONVENTIONS OF LANGUAGE USAGE

The Apostrophe

1. To form the possessive of a name, the simplest (and perhaps the best) thing to do is to add *'s*, even when the name already ends with a sibilant (*-s, -cks, -x, -z*). Thus:

El Greco's colors

Rubens's models

Velázquez's subjects

Augustus John's sketches (his last name is *John*)

Jasper Johns's recent work (his last name is *Johns*)

But some authorities say that to make the possessive for names ending in a sibilant, only an apostrophe is added (without the additional *s*)— Velázquez would become Velázquez', and Moses would become Moses'—unless (1) the name is a monosyllable (e.g., Jasper Johns would still become Johns's) or (2) the sibilant is followed by a final *e* (Horace would still become Horace's). Note that despite the final *s* in Degas and the final *x* in Delacroix, these names do not end in a sibilant (the letters are not pronounced), and so the possessive must be made by adding *'s*.

2. Don't add *'s* to the title of a work to make a possessive; the addition can't be italicized (underlined) since it is not part of the title, and it looks odd attached to an italicized word. So, not "*The Sower*'s colors" and not "*The Sower's* colors"; rather, "the colors of *The Sower*."

3. Don't confuse *its* and *it's*. The first is a possessive pronoun ("Its colors have faded"); the second is a contraction of *it is* ("It's important to realize that most early landscapes were painted indoors"). You'll have no trouble if you remember that *its*, like other possessive pronouns such as *ours, his, hers, theirs*, does *not* use an apostrophe.

Capitalization

Most writers capitalize names of sharply limited periods (e.g., Pre-Columbian, Early Christian, Romanesque, High Renaissance, Rococo) and the names of movements (e.g., Impressionism, Minimalism, Symbolism). Most writers do not capitalize "classic" and "romantic"—but even if you do capitalize Romantic when it refers to a movement ("Delacroix was a Romantic painter"), note that you should not capitalize it when it is used in other senses ("There is something romantic about ruined temples").

Many writers capitalize the chief events of the Bible, such as the Creation, the Fall, the Annunciation, and the Crucifixion, and also mythological events, such as the Rape of Ganymede and the Judgment of Paris. Again, be consistent.

On capitalization in titles, see pages 221 and 234.

The Dash

Type a dash by typing two hyphens (--) without hitting the space bar before, between, or after. Do not confuse the dash with the hyphen. Here is an example of the dash:

> New York--not Paris--is the center of the art world today.

The Hyphen

1. **Use a hyphen to divide a word at the end of a line.** Because words may be divided only as indicated by a dictionary, it is easier to end the line with the last complete word you can type than to keep reaching for a dictionary. But here are some principles governing the division of words at the end of a line:

- Never hyphenate words of one syllable, such as *called, wrote, doubt, through.*
- Never hyphenate so that a single letter stands alone: *a-lone, hair-y.*
- If a word already has a hyphen, divide it at the hyphen: *anti-intellectual.*
- Divide prefixes and suffixes from the root: *pro-vide; paint-ing.*
- Divide between syllables. Most words with double consonants should be hyphenated between the double letters: *bal-let, bal-loon.* If you aren't sure of the proper syllabification, check a dictionary.

Notice that when hyphenating, you do not hit the space bar before or after hitting the hyphen.

2. **Use a hyphen to compound adjectives into a single visual unit:** *twentieth-century architects, mid-century architecture* (but "She was born in the twentieth century").

Foreign Words and Quotations in Foreign Languages

1. **Underline (to indicate italics) foreign words that are not part of the English language.** Examples: *sfumato* (an Italian word for a

"blurred outline"), *pai-miao* (Chinese for "fine-line work"). But such words as chiaroscuro and Ming are not italicized because, as their presence in English dictionaries indicates, they have been accepted into English vocabulary. Foreign names are discussed below.

2. **Do not underline a quotation (whether in quotation marks or set off) in a foreign language.** If you quote a full sentence or more in a foreign language, for instance a passage from a critic, do not underline or italicize it. A word about foreign quotations: If your paper is frankly aimed at specialists, you need not translate quotations from languages that your readers might be expected to know, but if it is aimed at a general audience, translate foreign quotations, either immediately below the original or in a footnote.

3. On **translating the titles of works of art,** and on **capitalizing the titles of foreign books,** see "Titles," page 234.

Names

1. **Arabic names** require caution in alphabetizing in bibliographies. For example, family names beginning with *abu-* are alphabetized under this element but those beginning with *al-* are alphabetized under the next element. For details, consult *The Chicago Manual of Style*, 14th edition.

2. **Asian names,** in the original language, customarily put the family name first. Thus, the Japanese architect who in the West is known as Kenzo Tange, and whose name is alphabetized under Tange, in Japan is known as Tange Kenzo, and he is addressed as Tange-san (Mr. Tange). Similarly, the Korean-American artist whom we call Nam June Paik is, in Korea, called Paik Nam June, Paik being the family name. Until recently, Chinese, Japanese, and Korean names when given in English were given in the Western order, with the family name last ("Kenzo Tange's influence is still great"), but today, in an effort to avoid eurocentrism, there is a notable tendency to use the original sequence, as in "Tange Kenzo's influence is still great." In a bibliography the name of course is alphabetized under *T,* the first letter of the family name. One last (complicated) example: Wu Hung, a historian of Chinese art at the University of Chicago, is Professor Wu, because Wu is his family name. But on his books—even those published in English—his name appears as Wu Hung, i.e., the name is given in the Chinese style. An American student, innocent of Chinese, might think, wrongly, that in a bibliography this author would be alphabetized under *H.* If in preparing a bibliography you

are in doubt about which is the family name, ask someone who is likely to know.

3. **Dutch *van*,** as in Vincent van Gogh, is never capitalized by the Dutch except when it begins a sentence; in American usage, however, it is acceptable (but not necessary) to capitalize it when it appears without the first name, as in "The paintings of Van Gogh." But: "Vincent van Gogh's paintings." Names with *van* are commonly alphabetized under the last name, for example under *G* for Gogh.

4. **French *de*** is not used (De Gaulle was an exception) when the first name is not given. Thus, the last name of Georges de La Tour is La Tour, *not* de La Tour. But when *de* is combined with the definite article, into *des* or *du,* it is given even without the first name. *La* and *Le* are used even when the first name is not given, as in Le Nain.

5. **Spanish *de*** is not used without the first name, but if it is combined with *el* into *del,* it is given even without the first name.

6. **Names of deceased persons** are never prefaced with Mr., Miss, Mrs., or Ms.—even in an attempt at humor.

7. **Names of women** are not prefaced by Miss, Ms., or Mrs.; treat them like men's names—that is, give them no title.

8. **First names alone** are used for many writers and artists of the Middle Ages and Renaissance (examples: Dante, for Dante Alighieri; Michelangelo, for Michelangelo Buonarroti; Piero, for Piero della Francesca; Rogier, for Rogier van der Wyeden), and usually these are even alphabetized under the first name. Leonardo da Vinci, born near the town of Vinci, is Leonardo, never da Vinci. But do not adopt a chatty familiarity with later people: Picasso, not Pablo. *Exception:* Because van Gogh often signed his pictures "Vincent," some writers call him Vincent.

Avoiding Sexist Language

1. Traditionally, the male pronouns *he* and *his* have been used generically—that is, to refer to both men and women ("An architect should maintain his independence"). But this use of *he* and *his* offends many readers and is no longer acceptable. Common ways to avoid this type of sexist language are to use *he or she, she or he, s/he, (s)he, he/she, she/he, his or her,* or *her or his.* Some writers shift from masculine to feminine forms in alternating sentences or alternating paragraphs, and a few writers regularly use *she* and *her* in place of the generic *he* and *his* in order to make a sociopolitical point. But these expressions usually call too much

attention to themselves. Consider, for example, this grotesque sentence from an article in *Art History* 15:4 (1992), page 545:

> What some music also does (and particularly Wagner) is draw the attentive listener into it, so that s/he finds him- or herself in a close dialectical engagement with something which seems like his- or herself in character but which is neither quite this, nor yet quite alien.

The writer is trying to say something about music, but his attempts to avoid sexist language—to show that he belongs to the correct political party—by writing "s/he . . . him- or herself . . . his- or herself" are so awkward and so conspicuous that the reader notices only them, not the real point of the sentence.

There are other, more effective ways of avoiding sexist writing. Often you can substitute the plural form. For instance, instead of

> An architect should maintain his independence.

you can write

> Architects should maintain their independence.

Or you can recast the sentence to eliminate the possessive pronoun:

> An architect should be independent.

2. Think twice before you use *man* or *mankind* in such expressions as "man's art" or "the greatness of mankind." Consider using instead such words as *human being, person, people, humanity,* and *we.* (Examples: "Human beings need art," or "Humankind needs art," or "We need art," instead of "Man needs art.")

3. *Layman, craftsman,* and similar words should be replaced with such gender-neutral substitutes as *layperson* or *unspecialized people,* and (for *craftsman*) *craftsperson, craftworker,* or probably better, *artisan* or a more specific term such as *fabric artist.* Unfortunately, there seems to be no adequate synonym for *craftsmanship,* although in some contexts *technique* or *technical skill* will do nicely.

4. Just as you would not describe van Gogh as "the male painter" without good reason, you should not use such expressions as "woman painter" or "female sculptor" unless the context requires them.

5. *Reminder:* In the minds of some readers, words such as *potent* and *seminal* imply masculine values. (See page 160.)

Avoiding Eurocentric Language

The art history that most Americans are likely to encounter has been written chiefly by persons of English or European origin. Until recently such writing saw things from a European point of view and tended to assume the preeminence of European culture. Certain English words that convey this assumption of European superiority, such as *primitive* applied to African or Oceanic art, now are widely recognized as naive. Some words, however, are less widely recognized as outdated and offensive. For instance, the people whom Caucasians long have called *Eskimo* prefer to be called *Inuit,* and there is no reason why we should not honor their preference.

Asian; Oriental *Asian,* as a noun and as an adjective, is preferable. *Oriental* (from *oriens,* "rising sun," "east") is in disfavor because it implies a Eurocentric view (i.e., things "oriental" were east of the European colonial powers). Similarly, **Near East** (the countries of the eastern Mediterranean, Southwest Asia, the Arabian Peninsula, and sometimes northern Africa), **Middle East** (variously defined, but usually the area in Asia and Africa between and including Libya in the west, Pakistan in the east, Turkey in the north, and the Arabian Peninsula in the south), and **Far East** (China, Vietnam, North and South Korea, and Japan, or these and all other Asian lands east of Afghanistan) are terms based on a Eurocentric view. No brief substitutes have been agreed on for *Near East* and *Middle East,* but *East Asia* is now regarded as preferable to *Far East.* On Asian names, see pages 226–27, 260.

Eskimo; Inuit *Eskimo* (from the Algonquin for "eaters of raw meat") is a name given by the French to those native people of Canada who call themselves *Inuit* (singular: *Inuk*). The Inuit regard *Eskimo* as pejorative, and their preference is now officially recognized in Canada. *Eskimo* is still commonly used, however, as a general label for the group of North American Arctic cultures that includes, among others, the Yupik in Alaska, the Inuit in Canada, and the Greenlanders in Greenland.

Far East See *Asian.*

Hispanic The word—derived from *Hispania,* the Latin name for Spain—is widely used to designate not only persons from Spain but also members of the various Spanish-speaking communities living in the United States—Puerto Ricans, Cuban Americans, and persons from South America and Central America (including Mexican Americans, sometimes called Chicanos). Some members of these communities, however, strongly object to the term, arguing that it overemphasizes their European heritage and ignores the Indian and African heritages of many of the people it claims to describe. The same has been said of *Latina* and *Latino,* but these terms are more widely accepted within the communities, probably because the words are Spanish rather than English and therefore do not imply assimilation to Anglo culture. Further, *Latina* and *Latino* denote only persons of Central and South American descent, whereas *Hispanic* includes persons from Spain. Moreover, many people believe that the differences among Spanish-speaking groups from various countries are so great as to make *Hispanic* (like *Latino* and *Latina*) a reductive, almost meaningless label. Polls indicate that most persons in the United States who trace their origin to a Spanish-speaking country prefer to identify themselves as *Cuban, Mexican* (or *Chicano*), *Peruvian, Puerto Rican,* and so on, rather than as *Hispanic.*

Indian; Native American When Columbus encountered the Caribs in 1492, he thought he had reached India and therefore called them *Indios* (Indians). Later, efforts to distinguish the peoples of the Western Hemisphere produced the terms *American Indian, Amerindian,* and *Amerind.* More recently *Native American* has been used, but of course the people who met the European newcomers were themselves descended from persons who had emigrated from eastern Asia in ancient times, and in any case *American* is a word derived from the name of an Italian. On the other hand, anyone born in America, regardless of ethnicity, is a native American. Further, many Native Americans (in the new, restricted sense) continue to speak of themselves as Indians (e.g., members of the Navaho Indian Nation), thereby making the use of that word acceptable. Although *Indian* is acceptable, use the name of the specific group, such as Iroquois or Navaho, whenever possible. In Canada, however, the accepted terms now are *First Nations People* and *First Nations Canadians,* although some of these people call themselves Indians. The words *tribe* and *clan* are yielding to *nation* and *people* (e.g., the Zuni people), but in Canada *band* is widely used. One other point: All of the aboriginal peoples of Canada and Alaska can be called Native Americans, but some of them (e.g., the Inuit and the Aleut) cannot be called Indians.

Inuit; Eskimo See *Eskimo.*
Latina/Latino See *Hispanic.*
Native American See *Indian.*
New World; Western Hemisphere Although half of the earth—comprising North America, Mexico, Central America, and South America—in the late sixteenth century was new to Europeans, it was not new to the people who lived in this half of the world. The term *Western Hemisphere* is preferred to *New World.*
Primitive Derived ultimately from the Latin *primus,* meaning "first," *primitive* was widely used by anthropologists in the late nineteenth and early twentieth centuries with reference to nonliterate, nonwhite societies, e.g., in Africa south of the Sahara, in Oceania, and in pre-Columbian America. These societies were thought to be still in the first stages of an evolutionary process that culminates in "civilization," whose finest flowering was believed to be white industrial society. Even if "primitive" societies were regarded as having certain virtues, for instance "naturalness" or "spontaneity," these virtues were viewed with some condescension—the virtues were usually regarded as those of children—and present members of the society were usually thought to have lost them. Today virtually all anthropologists agree that the word *primitive* is misleading because it implies not only that the products of a "primitive" society (art, myths, and so on) are crude and simple, but also that such a society does not have a long, evolving history. **Tribal** is sometimes used as a substitute, but this word too contains condescending Eurocentric implications. (The journalists who in newspapers write of "tribal conflicts" in Africa would never speak of "tribal conflicts" in Europe.) *Aboriginal* and *non-Western* are sometimes used for what was once called "primitive art," but *aboriginal* gives off a condescending Eurocentric whiff, and *non-Western* inadvertently includes Asian art, for instance Ming porcelains. When speaking of the art of individual cultures it probably is best to use their names (e.g., "Olmec masks," "Yoruba sculpture," "Benin bronzes").
No term has emerged that can usefully and inoffensively be applied when speaking across cultures—and many people would argue that this is a good thing, since any term would necessarily make false connections among diverse, independent cultures.
Primitive is used also in two other senses: (1) to refer to the early stages of a particular school of painters—especially the Netherlandish painters of the late fourteenth and fifteenth centuries, and the Italian

painters between Giotto (born ca. 1267) and Raphael (born 1483)—with the false assumption that these early artists were trying to achieve the illusionistic representation that their successors achieved; and (2) to refer to works by artists untrained in formal academies, such as "Grandma Moses" (1860–1961) from upstate New York, who was regarded as preserving a naive, uncorrupted, childlike, charming vision. Paintings by these "primitives" are usually bright, detailed, and flat, with a strong emphasis on design. **Folk art** and **naive art** have long been substitute terms for this last sort of "primitive" art, but **vernacular art** now seems to be the most common term. These terms usually include the makers of functional objects such as baskets, quilts, and toys. If, however, the work is not functional and is highly distinctive, evidently the product of an individual psyche and not part of a recognizable tradition of art, it may be called **art brut** (French: "raw art," a term coined in 1945 by Jean Dubuffet), **visionary art,** or **outsider art.** The terms *art brut, visionary art,* and *outsider art* are especially used with reference to work produced by people who have had little or no formal training in art and who (unlike traditional folk artists) are largely isolated from the common culture, for instance the insane, religious visionaries, prisoners (see Phyllis Kornfeld, *Cellblock Visions,* 1997), and recluses. The works of these self-taught artists are usually paintings (often not on canvas but on cardboard, handkerchiefs, and other nontraditional surfaces), carvings, or assemblies of what most people regard as junk. For a review of five books on the topic, see N. F. Karlins in *Art Journal* 56:4 (Winter 1997), pages 93–97.

Tribal See *Primitive.*

Spelling

If you are a weak speller, ask a friend to take a look at your paper. If you have written the paper on a word processor, use the spell checker if there is one, but remember that the spell checker tells you only if a word is not in its dictionary. It does *not* tell you that you erred in using *their* where *there* is called for.

Experience has shown that the following words are commonly misspelled in papers on art. If the spelling of one of these words strikes you as odd, memorize it.

altar (noun)	dominant	shepherd
alter (verb)	exaggerate	silhouette
background	independent	spatial
connoisseur	parallel	subtly
contrapposto	prominent	symmetry
Crucifixion	recede	vertical (*not* verticle)
deity (*not* diety)	referring	watercolor (no hyphen)
dimension	separate	

Be careful to distinguish the following:

affect, effect *Affect* is usually a verb, meaning (1) "to influence, to produce an effect, to impress," as in "These pictures greatly affected the history of painting," or (2) "to pretend, to put on," as in "He affected to enjoy the exhibition." Psychologists use it as a noun for "feeling" ("The patient experienced no affect"), and we best leave this word to psychologists. *Effect,* as a verb, means "to bring about" ("The workers effected the rescue in less than an hour"). As a noun, *effect* means "result" ("The effect of his work was negligible").

capital, capitol A *capital* is the uppermost member of a column or pilaster; it also refers to accumulated wealth, or to a city serving as a seat of government. A *capitol* is a building in which legislators meet, or a group of buildings in which the functions of government are performed.

eminent, immanent, imminent *Eminent,* "noted, famous"; *immanent,* "remaining within, intrinsic"; *imminent,* "likely to occur soon, impending."

its, it's *Its* is a possessive ("Its origin is unknown"); *it's* is short for *It is* ("It's an early portrait"). You won't confuse these two words if you remember that possessive pronouns (*his, her, my, yours,* etc.) never take an apostrophe.

lay, lie To *lay* means "to put, to set, to cause to rest" ("Lay the glass on the print"). To *lie* means "to recline" ("Venus lies on a couch").

loose, lose *Loose* is an adjective ("The nails in the frame are loose"); *lose* is a verb ("Don't lose the nails").

precede, proceed To *precede* is to come before in time: to *proceed* is to go onward.

principal, principle *Principal* as an adjective means "leading," "chief"; as a noun it means a leader (and, in finance, wealth). *Principle* is only a noun, "a basic truth," "a rule," "an assumption."

Titles

1. **On the form of the title** of a manuscript essay, see pages 221–22.

2. **On underlining titles** of works of art, see the next section, "Italics and Underlining."

3. Some works of art are regularly given with their **titles in foreign languages** (Goya's *Los Caprichos,* the Limbourg Brothers' *Les Très Riches Heures du Duc de Berry,* Picasso's *Les Demoiselles d'Avignon*), and some works are given in a curious mixture of tongues (Cézanne's *Mont Sainte-Victoire Seen from Bibémus Quarry*), but the vast majority are given with English titles: Picasso's *The Old Guitarist,* Cézanne's *Bathers,* Millet's *The Gleaners.* In most cases, then, it seems pretentious to use the original title.

4. **Capitalization in foreign languages** is not the same as in English.

> *French:* When you give a title—of a book, essay, or work of art—in French, capitalize the first word and all proper nouns. If the first word is an article, capitalize also the first noun and any adjective that precedes it. Examples: *Le Déjeuner sur l'herbe; Les Très Riches Heures du Duc de Berry.*

> *German:* Follow German usage; for example, capitalize the pronoun *Sie* ("you"), but do not capitalize words that are not normally capitalized in German.

> *Italian:* Capitalize only the first word and the names of people and places.

Italics and Underlining

1. **Use italics or underlining for titles of works of art, other than architecture:** Michelangelo's *David,* van Gogh's *Sunflowers;* but the Empire State Building, the Brooklyn Bridge, the Palazzo Vecchio.

2. **Italicize or underline titles of journals and books** other than holy works: *Art Journal, Art and Illusion, The Odyssey;* but Genesis, the Bible, the New Testament, the Koran (or, now preferred, the Quran). For further details about biblical citations, see page 238.

3. **Italicize titles of art exhibitions.**

4. **Italicize or underline foreign words,** but use roman type for quotations from foreign languages (see page 226).

QUOTATIONS AND QUOTATION MARKS

If you are writing a research paper, you will need to include quotations, almost surely from scholars who have worked on the topic, and possibly from documents such as letters or treatises written by the artist or by the artist's contemporaries. But even in a short analysis, based chiefly on looking steadily at the object, you may want to quote a source or two—for example, your textbook. The following guidelines tell you how to give quotations and how to cite your sources—but remember, a good paper is not a bundle of quotations.

    1. **Be sparing in your use of quotations.** Use quotations as evidence, not as padding. If the exact wording of the original is crucial, or especially effective, quote it directly, but if it is not, don't bore the reader with material that can be effectively reduced either by summarizing or by cutting. If you cut, indicate ellipses as explained in point 5.

    2. **Identify the speaker or writer of the quotation,** so that readers are not left with a sense of uncertainty. Usually this identification precedes the quoted material (e.g., you write something like "Rosalind E. Krauss argues" or "Kenneth Clark long ago pointed out"), in accordance with the principle of letting readers know where they are going. But occasionally the identification may follow the quotation, especially if it will prove something of a pleasant surprise. For example, in a discussion of Jackson Pollock's art, you might quote a hostile comment on one of the paintings and then reveal that Pollock himself was the speaker. (Further suggestions about leading into quotations are given on page 218.)

    3. **Distinguish between short and long quotations,** and treat each appropriately. *Short quotations* (usually defined as fewer than five lines of prose) are enclosed within quotation marks and run into the text (rather than set off, without quotation marks).

> Michael Levey points out that "Alexander singled out Lysippus to be his favorite sculptor, because he liked the character given him in Lysippus' busts." In making this point, Levey is not taking the familiar view that. . . .

A *long quotation* (usually five or more lines of typewritten prose) is *not* enclosed within quotation marks. To set it off instead, the usual practice is to triple-space before and after the quotation and single-space the quotation, indenting five spaces—ten spaces for the first line if the quotation begins with the opening of a paragraph. (Note: The suggestion that you single-space longer quotations seems reasonable but is at odds with various manuals that tell how to prepare a manuscript for publication. Such manuals usually say that material that is set off should be indented ten spaces and double-spaced. Find out if your instructor has a preference.)

Introduce a long quotation with an introductory phrase ("Jones argues that") or with a sentence ending with a colon ("Jones offers this argument:"). For an example of this use of a colon, see the lead-in to the quotation from Leo Steinberg on page 236.

4. **An embedded quotation** (i.e., a quotation embedded in a sentence of your own) must fit grammatically into the sentence of which it is a part. For example, suppose you want to use Zadkine's comment, "I do not believe that art must develop on national lines."

Incorrect

> Zadkine says that he "do not believe that art must develop on national lines."

Correct

> Zadkine says that he does "not believe that art must develop on national lines."

Correct

> Zadkine says, "I do not believe that art must develop on national lines."

Don't try to introduce a long quotation (say, more than a complete sentence) into the middle of one of your own sentences. It is almost impossible for the reader to come out of the quotation and to pick up the thread of your own sentence. It is better to lead into the long quotation with "Jones says"; and then, after the quotation, to begin a new sentence of your own.

5. **The quotation must be exact.** Any material that you add within a quotation must be in square brackets (not parentheses), thus:

> Pissarro, in a letter, expressed his belief that "the Japanese practiced this art [of using color to express ideas] as did the Chinese."

If you wish to omit material from within a quotation, indicate the ellipsis by three spaced periods. If a sentence ends in an omission, add a closed-up period and then three spaced periods to indicate the omission. The following example is based on a quotation from the sentences immediately before this one:

> The manual says that "if you . . . omit material from within a quotation, [you must] indicate the ellipsis. . . . If a sentence ends in an omission, add a closed-up period and then three spaced periods. . . ."

Notice that although material preceded "if you," an ellipsis is not needed to indicate the omission because "if you" began a sentence in the original. (Notice, too, that although in the original *if* was capitalized, in the quotation it is reduced to lowercase in order to fit into the sentence grammatically.) Customarily initial and terminal omissions are indicated only when they are part of the sentence you are quoting. Even such omissions need not be indicated when the quoted material is obviously incomplete—when, for instance, it is a word or phrase.

    6. **Punctuation is a bit tricky.** Commas and periods go *inside* the quotation marks; semicolons and colons go *outside* the marks.

    Question marks, exclamation points, and dashes go inside if they are part of the quotation, outside if they are your own. Compare the positions of the question marks in the two following sentences.

> The question Jacobs asked is this: "Why did perspective appear when it did?" Can we agree with Jacobs that "perspective appeared when scientific thinking required it"?

    7. **Use single quotation marks for material contained within a quotation** that itself is within quotation marks. In the following example, a student quotes William Jordy (the quotation from Jordy is enclosed within double quotation marks), who himself quoted Frank Lloyd Wright (the quotation within Jordy's quotation is enclosed within single quotation marks):

> William H. Jordy believes that to appreciate Wright's Guggenheim Museum one must climb up it, but he recognizes that "Wright . . . recommended that one take the elevator and circle downward. 'The elevator is doing the lifting,' as he put it, 'the visitor the drifting from alcove to alcove.'"

    8. **Use quotation marks around titles of short works**—that is, for titles of chapters in books and for stories, essays, short poems, songs, lectures, and speeches. Titles of unpublished works, even book-length dissertations, are also enclosed in quotation marks. But underline—to indicate *italics*—titles of pamphlets, periodicals, and books. Underline also

titles of films, radio and television programs, ballets and operas, and works of art except architecture. Thus: Michelangelo's *David*, Picasso's *Guernica*, Frank Lloyd Wright's Guggenheim Museum.

Exception: Titles of sacred writings (e.g., the Old Testament, the Hebrew Bible, the Bible, Genesis, Acts, the Gospels, the Quran) are not underlined, not italicized, and not enclosed within quotation marks. Incidentally, it is becoming customary to speak of the Hebrew Bible or of the Hebrew Scriptures, rather than of the Old Testament, and some writers speak of the Christian Scriptures rather than of the New Testament. The objection to the term "Old Testament" is based on the idea that the Hebrew writings are implicitly diminished when they are regarded as "Old" writings that are replaced by "New" ones. Although "Hebrew Bible" is not entirely accurate since some parts of the Scriptures of Judaism are written not in Hebrew but in Aramaic, it is the preferred term by many Jews and by those Christians who are aware of the issue.

To cite a book of the Bible with chapter and verse, give the name of the book, then a space, then an arabic numeral for the chapter, a period, and an arabic numeral (*not* preceded by a space) for the verse, thus: Exodus 20.14–15. (The older method of citation, with a small roman numeral for the chapter and an arabic numeral for the verse, is no longer common.) Standard abbreviations for the books of the Bible (for example, 2 Cor. for 2 Corinthians) are permissible in citations.

ACKNOWLEDGING SOURCES

Borrowing Without Plagiarizing

You must acknowledge your indebtedness for material when

1. You quote directly from a work.
2. You paraphrase or summarize someone's words (the words of the paraphrase or summary are your own, but the points are not, and neither, probably, is the structure of the sentences).
3. You appropriate an idea that is not common knowledge.

Let's suppose you want to make use of William Bascom's comment on the earliest responses of Europeans to African art:

> The first examples of African art to gain public attention were the bronzes and ivories which were brought back to Europe after the sack of Benin by a British military expedition in 1897. The superb technol-

ogy of the Benin bronzes won the praise of experts like Felix von Luschan who wrote in 1899, "Cellini himself could not have made better casts, nor anyone else before or since to the present day." Moreover, their relatively realistic treatment of human features conformed to the prevailing European aesthetic standards. Because of their naturalism and technical excellence, it was at first maintained that they had been produced by Europeans—a view that was still current when the even more realistic bronze heads were discovered at Ife in 1912. The subsequent discovery of new evidence has caused the complete abandonment of this theory of European origins of the bronzes of Benin and Ife, both of which are cities in Nigeria.

—William Bascom, *African Art in Cultural Perspective* (1973), 4.

1. **Acknowledging a direct quotation.** You may want to use some or all of Bascom's words, in which case you will write something like this:

> As William Bascom says, when Europeans first encountered Benin and Ife works of art in the late nineteenth century, they thought that Europeans had produced them, but the discovery of new evidence "caused the complete abandonment of this theory of European origins of the bronzes of Benin and Ife, both of which are cities in Nigeria."[1]

Notice that the digit, indicating a footnote, is raised, and that it follows the period and the quotation mark. (The form of footnotes is specified on pages 245–50.)

2. **Acknowledging a paraphrase or summary.** Summaries (abridgments) are usually superior to paraphrases (rewordings, of approximately the same length as the original) because summaries are briefer, but occasionally you may find that you cannot abridge a passage in your source and yet you don't want to quote it word for word—perhaps because it is too technical or because it is poorly written. Even though you are changing some or all of the words, you must give credit to the source because the idea is not yours, nor, probably, is the sequence of the presentation. Here is an example:

Summary

> William Bascom, in <u>African Art</u>, points out that the first examples of African art--Benin bronzes and ivories--brought to Europe were thought by Europeans to be of European origin, because of their naturalism and their technical excellence, but evidence was later discovered that caused this theory to be abandoned.

> ## ✒ A RULE FOR WRITERS:
>
> Acknowledge your sources
>
> 1. if you quote directly and put the quoted words in quotation marks
> 2. if you summarize or paraphrase someone's material, even though you do not retain one word of your source
> 3. if you borrow a distinctive idea, even though the words and the concrete application are your own

Not to give Bascom credit is to plagiarize, even though the words are yours. The offense is just as serious as not acknowledging a direct quotation. And, of course, if you say something like what is given in the following example and you do not give credit, you are also plagiarizing, even though almost all of the words are your own.

Plagiarized Summary

> The earliest examples of African art to become widely known in Europe were bronzes and ivories that were brought to Europe in 1897. These works were thought to be of European origin, and one expert said that Cellini could not have done better work. Their technical excellence, as well as their realism, fulfilled the European standards of the day. The later discovery of new evidence at Benin and Ife, both in Nigeria, refuted this belief.

It is pointless to offer this sort of rewording: If there is a point, it is to conceal the source and to take credit for thinking that is not your own.

3. **Acknowledging an idea.** Let us say that you have read an essay in which Seymour Slive argues that many Dutch still lifes have a moral significance. If this strikes you as a new idea and you adopt it in an essay—even though you set it forth entirely in your own words and with examples not offered by Slive—you should acknowledge your debt to Slive. Not to acknowledge such borrowing is plagiarism. Your readers will not think the less of you for naming your source; rather, they will be grateful to you for telling them about an interesting writer.

Similarly, if in one of your courses an instructor makes a point that you do not encounter in your reading and that therefore probably is not common knowledge (*common knowledge* will be defined in the next section), cite the instructor, the date, and the institution where the lecture was delivered.

Caution: Information derived from the Internet must be properly cited. Probably it is best if you do *not* download it into your own working text. Rather, create a separate document file so that later you will recognize it as material that is not your own.

Fair Use of Common Knowledge

When in doubt as to whether to give credit (either in a footnote or merely in an introductory phrase such as "William Bascom says"), give credit. As you begin to read widely in your field or subject, you will develop a sense of what is considered common knowledge. Unsurprising definitions in a dictionary can be considered common knowledge, so there is no need to say "According to Webster, a mural is a picture or decoration applied to a wall or ceiling." (That's weak in three ways: It's unnecessary, it's uninteresting, and it's inexact, since "Webster" appears in the titles of several dictionaries, some good and some bad.)

Similarly, the date of Picasso's death can be considered common knowledge. Few can give it when asked, but it can be found in many sources, and no one need get the credit for providing you with the date. Again, if you simply *know*, from your reading of Freud, that Freud was interested in art, you need not cite a specific source for an assertion to that effect, but if you know only because some commentator on Freud said so, and you have no idea whether the fact is well known or not, you should give credit to the source that gave you the information. Not to give credit—for ideas as well as for quoted words—is to plagiarize.

With matters of interpretation the line is less clear. For instance, almost all persons who have published discussions of van Gogh's *The Potato Eaters* have commented on its religious implications or resonance. In 1950 Meyer Schapiro wrote, "The table is their altar . . . and the food a sacrament. . . ." In 1971 Linda Nochlin wrote that the picture is an "overtly expressive embodiment of the sacred," and in 1984 Robert Rosenblum commented on the "almost ritualistic sobriety that seems inherited from sacred prototypes." If you got this idea from one source, cite the source, but if in your research you encountered it in several places, it will be enough if you say something like, "The sacramental quality of the picture has been widely noted." You need not cite half a dozen references—though you may wish to add, "first by," or "most recently by," or some such thing, in order to lend a bit of authority to your paper.

"But How Else Can I Put It?"

If you have just learned—say, from an encyclopedia—something that you sense is common knowledge, you may wonder, "How can I change into my own words the simple, clear words that this source uses in setting forth this simple fact?" For example, if before writing about Rosa Bonheur's painting of Buffalo Bill (he took his Wild West show to France), you look him up in the *Encyclopaedia Britannica,* you will find this statement about Buffalo Bill (William F. Cody): "In 1883 Cody organized his first Wild West exhibition." You cannot use this statement as your own, word for word, without feeling uneasy. But to put in quotation marks such a routine statement of what can be considered common knowledge, and to cite a source for it, seems pretentious. After all, the *Encyclopedia Americana* says much the same thing in the same routine way: "In 1883, . . . Cody organized Buffalo Bill's Wild West." It may be that the word "organized" is simply the most obvious and the best word, and perhaps you will end up using it. Certainly to change "Cody organized" into "Cody presided over the organization of" or "Cody assembled" or some such thing, in an effort to avoid plagiarizing, would be to make a change for the worse and still to be guilty of plagiarism. But you won't get yourself into this mess of wondering whether to change clear, simple wording into awkward wording if in the first place, when you take notes, you summarize your sources, thus: "1883: organized Wild West," or "first Wild West: 1883." Later (even if only thirty minutes later), when drafting your paper, if you turn this nugget—probably combined with others—into the best sentence you can, you will not be in danger of plagiarizing, even if the word "organized" turns up in your sentence.

Notice that *taking notes* is part of the trick; this is not the same thing as copying or photocopying. Photocopying machines are great conveniences but they also make it easy for us not to think; we later may confuse a photocopy of an article with a thoughtful response to an article. The copy is at hand, a few words underlined, and we use the underlined material with the mistaken belief that we have absorbed it.

If you take notes thoughtfully, rather than make copies mindlessly, you will probably be safe. Of course, you may want to say somewhere in your paper that all your facts are drawn from such-and-such a source, but you offer this statement not to avoid charges of plagiarism but for three other reasons: to add authority to your paper, to give respectful credit to those who have helped you, and to protect yourself in case your source contains errors of fact.

DOCUMENTATION

As a student, you are a member of a community of writers who value not only careful scholarship and good writing but also full and accurate documentation. Various academic disciplines have various systems of documentation—the footnote form of professors of literature differs from the footnote form of professors of sociology. Even within a discipline there may be alternative styles. For instance, the College Art Association publishes *Art Bulletin* and *Art Journal*, but the systems of documentation used by these journals are not identical, and most university presses use a system very different from both of these journals even when they publish books about art.

In pages 245–53 you will find the principles set forth in *The Chicago Manual of Style*, 14th edition (1993), the guide followed by most university presses and by many scholarly journals—but not by either of the journals published by the College Art Association. If your instructor wants you to use *Art Bulletin* style, you will find it on pages 253–63.

For ease of reference when you are preparing a paper, we have tinted the outside margins of the pages of these two style guides; pages tinted on the upper half give Chicago style; pages tinted on the lower half give *Art Bulletin* style. Whichever style you use, be consistent.

FOOTNOTES AND ENDNOTES

Kinds of Notes

In speaking of kinds of notes, this paragraph does not distinguish between **footnotes,** which appear at the bottom of the page, and **endnotes,** which appear at the end of the essay; for simplicity, *footnote* will cover both terms. Rather, a distinction is made between (1) notes that give the sources of quotations, facts, and opinions used and (2) notes that give additional comment that would interrupt the flow of the argument in the body of the paper.

This second type perhaps requires a comment. You may wish to indicate that you are familiar with an opinion contrary to the one you are offering, but you may not wish to digress upon it during the course of your argument. A footnote lets you refer to it and indicate why you are not considering it. Or a footnote may contain statistical data that support your point but that would seem unnecessarily detailed and even tedious in the body of the paper. This kind of footnote, giving additional commentary, may have its place in research papers and senior theses, but even in such essays it should be used sparingly, and it rarely has a place in a short analytical essay. There are times when supporting details may be appropriately relegated to a footnote, but if the thing is worth saying, it is usually worth saying in the body of the paper. Don't get into the habit of affixing either trivia or miniature essays to the bottom of each page of an essay.

Footnote Numbers and Positions

Number the notes consecutively throughout the essay or chapter. Although some instructors allow students to group all the notes at the rear of the essay, most instructors—and surely all readers—believe that the best place for a note is at the foot of the appropriate page.

If you use a word processor, your software may do much of the job for you. It probably can automatically elevate the footnote number, and it can automatically print the note on the appropriate page. (For more on using a computer, see page 214.)

Footnote Style

To indicate that there is a footnote, put a raised arabic numeral (without a period and without parentheses) after the final punctuation of the sentence, unless clarity requires it earlier. In a sentence about Claude Monet, Camille Pissarro, and Alfred Sisley you may need a footnote for each and a corresponding numeral after each name instead of one numeral at the end of the sentence, but usually a single reference at the end will do. The single footnote might explain that Monet says such and such in a book entitled ———, Pissarro says such and such in a book entitled ———, and Sisley says such and such in a book entitled ———.

CHICAGO MANUAL STYLE

If you are using endnotes, begin the page with the heading "Notes" and then give the notes in numerical order. If you are placing the notes at the foot of the appropriate pages, double-space twice (i.e., skip four lines) at the bottom of the page before giving the first footnote. In either case, follow these principles:

- indent five spaces
- type the arabic number and a period
- skip one space and type the footnote, double-spacing it, beginning with a capital letter and putting a period at the end
- if the note runs more than one line, begin subsequent lines at the left margin
- begin each new note with the indentation of five spaces
- double-space between footnotes

Books

First Reference to a Book

> 1. Elizabeth ten Grotenhuis, <u>Japanese Mandalas: Representations of Sacred Geography</u> (Honolulu: University of Hawaii Press, 1999), 153.

Explanation:

- Give the author's name as it appears on the title page, *first name first.*
- You need not give a subtitle, but if you do give it (as in this example), separate it from the title with a colon and underline it.
- Give the name of the city of publication; if the city is not likely to be known, or if it can be confused with another city of the same name (Cambridge, Massachusetts, and Cambridge, England), add the name of the state or country, using an abbreviation.
- Give the page number (here, 153) after the comma that follows the closing parenthesis, with one space between the comma and the page number. Do *not* use "page" or "pages" or "p." or "pp."
- End the note with a period.

When you give the author's name in the body of the text—for instance in such a phrase as "Elizabeth ten Grotenhuis points out that"—do not repeat the name in the footnote. Merely begin with the title:

    2. Japanese Mandalas: Representations of Sacred Geography (Honolulu: University of Hawaii Press, 1999), 153.

A Revised Edition of a Book

    3. Rudolf Wittkower, Art and Architecture in Italy, 1600-1750, 3rd ed. (Harmondsworth, England: Penguin, 1973), 187.

A Book in More than One Volume

    4. Ronald Paulson, Hogarth: His Life, Art, and Times (New Haven, Conn.: Yale University Press, 1971), 2:161.

The reference here is to page 161 in volume 2. Abbreviations such as "vol." and "p." are *not* used.

A Book by More than One Author

    5. Romare Bearden and Harry Henderson, A History of African-American Artists from 1792 to the Present (New York: Pantheon, 1994), 612-13.

- The name of the second author, like that of the first, is given *first name first.*

- If there are more than three authors, give the full name of the first author and follow it with "and others" and a comma.

An Edited or Translated Book

    6. Ruth Magurn, ed. and trans., The Letters of Peter Paul Rubens (Cambridge, Mass.: Harvard University Press, 1955), 238.

    7. Helmut Brinker and Hiroshi Kanazawa, Zen Masters of Meditation in Images and Writings, trans. Andreas Leisinger (Zurich: Artibus Asiae, 1996), 129.

An Introduction or Foreword by Another Author
You may need to footnote a quotation from someone's introduction (say, Kenneth Clark's) to someone else's book (say, James Hall's). If in

your text you say, "As Kenneth Clark points out," the footnote will run thus:

> 8. Introduction to James Hall, <u>Dictionary of Subjects and Symbols in Art</u>, 2nd ed. (New York: Harper and Row, 1979), viii.

An Essay in a Collection of Essays by Various Authors

> 9. Charles Pellet, "Jewelers with Words," in <u>Islam and the Arab World</u>, ed. Bernard Lewis (New York: Knopf, 1976), 151.

As note 9 indicates, when you are quoting from an essay in an edited book,

- begin with the essayist (first name first) and the essay
- then give the title of the book and the name of the editor, first name first

References to Material Reprinted in a Book

Suppose you are using a book that consists of essays or chapters or pages by various authors, reprinted from earlier publications, and you want to quote a passage.

- If you have not given the author's name in the lead-in to your quotation, give it at the beginning of the footnote.
- Then give the title of the essay or of the original book.
- Then, if possible, give the place where this material originally appeared (you can usually find this information in the acknowledgments page of the book in hand or on the first page of the reprinted material).
- Then give the name of the title of the book you have in hand.
- Then give the editor of the collection, the place of publication, the publisher, the date, and the page number.

The monstrous but accurate footnote might run like this:

> 10. Henry Louis Gates, Jr., "The Face and Voice of Blackness," in <u>Facing History: The Black Image in American Art, 1710-1940</u>, ed. Guy McElroy (San Francisco: Bedford Art, 1990); rpt. in <u>Modern Art and Society</u>, ed. Maurice Berger (New York: HarperCollins, 1994), 53.

You have read Gates's essay, "The Face and Voice of Blackness," which was originally published in a collection (*Facing History*) edited by McElroy, but you did not read the essay in McElroy's collection. Rather, you read it in Berger's collection of reprinted essays, *Modern Art and Society*. You learned the name of McElroy's collection and the original date and place of publication from Berger's book, so you give this information, but your page reference is of course to the book that you are holding in your hand, page 53 of Berger's book.

Journals and Newspapers

An Article in a Journal with Continuous Pagination Throughout the Annual Volume

> 11. Anne H. van Buren, "Madame Cézanne's Fashions and the Dates of Her Portraits," Art Quarterly 29 (1966): 119.

The author's first name is given first; no month or season is given because there is only one page 119 in the entire volume; and abbreviations such as "vol." and "p." are *not* used.

An Article in a Journal that Paginates Each Issue Separately

> 12. Christine M. E. Guth, "Japan 1868-1945: Art, Architecture, and National Identity," Art Journal 55, no. 3 (1996): 17.

The issue number is given because for this quarterly journal there are, in any given year, four pages numbered 17.

An Article in a Popular Magazine

> 13. Henry Fairlie, "The Real Life of Women," New Republic, 26 August 1978, 18.

For popular weeklies and monthlies, give only the date (not the volume number), and do not enclose the date within parentheses.

A Book Review

If a book review has a title, treat the review as an article. If, however, the title is merely that of the book reviewed, or even if the review has a title but for clarity you wish to indicate that it is a review, the following form is commonly used:

> 14. Pepe Karmel, review of Off the Wall: Robert Rauschenberg and the Art World of Our Time, by Calvin Tomkins (Garden City, N.Y.: Doubleday, 1980), New Republic, 21 June 1980, 38.

A Newspaper
    The first example is for a signed article, the second for an unsigned one.

> 15. Bertha Brody, "Illegal Immigrant Sculptor Allowed to Stay," New York Times, 4 July 1994, 12.

> 16. "Portraits Stolen Again," Washington Post, 30 June 1995, 7.

Secondhand References

Let's assume you are reading a book (in this case, the fourth volume of a work by William Jordy) and the author quotes a passage (by Frank Lloyd Wright) that you want to quote in your essay. Your footnote should indicate both the original source if possible (i.e., not only Wright's name but also his book, place and year of publication, etc.), and then full information about the place where you found the quoted material:

> 17. Frank Lloyd Wright, The Solomon Guggenheim Museum (New York: Museum of Modern Art, 1960), 20; quoted in William H. Jordy, American Buildings and Their Architects (Garden City, New York: Anchor, 1976), 4:348.

If Jordy had merely given Wright's name and had quoted him but had not cited the source, of course you would be able to give only Wright's name and then the details about Jordy's book.

Subsequent References

When you quote a second or third or fourth time from the same work, use a short form in the subsequent notes. The most versatile form is simply the author's last name, an abbreviated title, and the page number:

> 18. Wittkower, Art, 38.

You can even dispense with the author's name if you have mentioned it in the sentence to which the footnote is keyed, and if the source is the one you have mentioned in the preceding note, you can use "Ibid." (an abbreviation of the Latin *ibidem,* "in the same place") and follow it with a comma and the page number.

> 19. Ibid., 159.

Although "Ibid." is Latin, customarily it is *not* italicized. If the page is identical with the page cited in the immediately preceding note, do not repeat the page number.

Interviews, Lectures, and Letters

20. Malcolm Rogers, Director, Museum of Fine Arts, Boston, interview by the author, Cambridge, Mass., 12 July 1998.

21. Howard Saretta, "Masterpieces from Africa," lecture at Tufts University, 13 May 1999.

22. Information in a letter to the author, from James Cahill, University of California, Berkeley, 17 March 1995.

Electronic Citations

See pages 208–10.

Bibliography (List of Works Cited)

A bibliography is a list of the works cited or, less often, a list of all relevant sources. (There is rarely much point in the second sort; if a particular book or article wasn't important enough to cite, why list it?) Normally a bibliography is given only in a long manuscript such as a research paper or a book, but instructors may require a bibliography even for a short paper if they wish to see at a glance the material that the student has used. In this case a heading such as "Works Cited" is less pretentious than "Bibliography."

Bibliographic Style

Because a bibliography is arranged alphabetically by author, the author's *last name is given first* in each entry. If a work is by more than one author, it is given under the first author's name; this author's last name is given first, but the other author's or authors' names follow the normal order of first name first. (See the entry under Rosenfield, page 251.)

Anonymous works are listed by title at the appropriate alphabetical place, giving the initial article, if any, but alphabetizing under the next word. Thus an anonymous essay entitled "A View of Leonardo" would retain the "A" but would be alphabetized under V for "View."

In typing an entry, use double-spacing. Begin flush with the lefthand margin; if the entry runs over the line, indent the subsequent lines of the entry five spaces. Double-space between entries.

A Book by One Author

> Caviness, Madeline Harrison. The Early Stained Glass of Canterbury Cathedral. Princeton: Princeton University Press, 1977.

An Exhibition Catalog

An exhibition catalog may be treated as a book, but some journals add "exh. cat." after the title of a catalog. The first example, a catalog that includes essays by several authors, gives the editor's name, which is specified on the title page. The second example is a catalog by a single author.

> Barnet, Peter, ed. Images in Ivory: Precious Objects of the Gothic Age, exh. cat. Detroit: Detroit Institute of Arts, 1997.

> Tinterow, Gary. Master Drawings by Picasso, exh. cat. Cambridge, Mass.: Fogg Art Museum, 1981.

A Book or Catalog by More Than One Author

> Rosenfield, John M., and Elizabeth ten Grotenhuis. Journey of the Three Jewels: Japanese Buddhist Paintings from Western Collections. New York: Asia Society, 1979.

Notice in this entry that although the book is alphabetized under the *last name* of the *first* author, the name of the second author is given in the ordinary way, first name first.

A Collection or Anthology

> Goldwater, Robert, and Marco Treves, eds. Artists on Art. New York: Pantheon, 1945.

This entry lists the collection alphabetically under the first editor's last name. Notice that the second editor's name is given first name first. A collection may be listed either under the editor's name or under the first word of the title.

An Essay in a Collection or Anthology

> Livingstone, Jane, and John Beardsley. "The Poetics and Politics of Hispanic Art: A New Perspective." In Exhibiting Cultures: The

> <u>Poetics and Politics of Museum Display</u>, ed. Ivan Karp and Steven D. Lavine, 104-20. Washington, D.C.: Smithsonian, 1991.

This entry lists an article by Livingstone and Beardsley (notice that the first author's name is given with the last name first, but the second author's name is given first name first) in a book called *Exhibiting Cultures*, edited by Karp and Lavine. The essay appears on pages 104–20.

Two or More Works by the Same Author

> Cahill, James. <u>Chinese Painting</u>. Geneva: Skira, 1960.
>
> ____. <u>Scholar Painters of Japan: The Nanga School</u>. New York: Asia House, 1972.

The horizontal line (eight units of underlining, followed by a period and then two spaces) indicates that the author (in this case James Cahill) is the same as in the previous item. Note also that multiple titles by the same author are arranged alphabetically (*Chinese* precedes *Scholar*).

An Introduction to a Book by Another Author

> Clark, Kenneth. Introduction to <u>Dictionary of Subjects and Symbols in Art</u>, by James Hall, 2nd ed. New York: Harper and Row, 1979.

This entry indicates that the student made use of the introduction rather than the main body of the book; if the body of the book were used, the book would be alphabetized under *H* for Hall, and the title would be followed by: Intro. Kenneth Clark.

An Edited Book

> Rossetti, Dante Gabriel. <u>Letters of Dante Gabriel Rossetti</u>. Edited by Oswald Doughty and J. R. Wahl. 4 vols. Oxford: Clarendon, 1965.

A Journal Article

> Mitchell, Dolores. "The 'New Woman' as Prometheus: Women Artists Depict Women Smoking." <u>Woman's Art Journal</u> 12, no. 1 (1991): 2-9.

Because this journal paginates each issue separately, the issue number must be given. For a journal that paginates issues continuously, give the year without the issue number.

A Newspaper Article

> "Museum Discovers Fake." <u>New York Times</u>, 21 January 1980, D29.

Romero, Maria. "New Sculpture Unveiled." <u>Washington Post</u>, 18 March
     1980, 6.

Because the first of these newspaper articles is unsigned, it is alpha-
betined under the first word of the title; because the second is signed, it
is alphabetized under the author's last name.

A Book Review

Gevisser, Mark. Review of <u>Art of the South African Townships</u>, by Gavin
     Younger. <u>Art in America</u> 77, no. 7 (1989): 35-39.

This journal paginates each issue separately, so the issue number must be
given as well as the year.

Electronic Sources
See pages 208–10.

ART BULLETIN STYLE*

The following material is a verbatim reprint of the *Art Bulletin Style
Guide,* except that it omits material relevant only to contributors to *Art
Bulletin,* such as instructions concerning the maximum length of manuscripts, the address to which manuscripts are to be sent, and the nature of
the author's responsibility to pay costs incurred in production.

Preparing the Manuscript
     For general questions of style, see *Chicago Manual of Style,*
14th ed. (University of Chicago Press), and *MLA Handbook for
Writers of Research Papers,* 3rd ed. (Modern Language Association, New York). For spelling, refer to *Webster's Third New International Dictionary or Webster's New Collegiate Dictionary.*
     The following are important points quite often overlooked
by authors. Special attention should be paid to differences from
the published style guides, in particular the style for footnotes,
bibliography (Frequently Cited Sources), and captions.
     Double space *all* copy: text, quotations, footnotes, captions,
Frequently Cited Sources, abstract of the article, author's biographical statement.

---

*Reprinted by permission of the College Art Association.

Footnotes should be numbered consecutively and typed double spaced on separate pages at the end of the article. Footnotes should not appear on the manuscript page with the text. Captions and Frequently Cited Sources should also be on separate pages, double spaced.

Leave a margin of one-and-a-half inches at the left and one inch at the top, bottom, and right. Do not break words (hyphenate) at ends of lines. *Do not* justify the right-hand margin. Use *italic* type or Roman type *underscored* for words to be set in italics.

Pagination should be consecutive. Do not use "page 11a" for a last-minute addition between 11 and 12, but renumber the succeeding pages. (The same applies to footnote numbers and other sequences.)

Text (Including Text of Footnotes)

1. Titles

Titles of works of art, like those of books, periodicals, exhibitions, etc., should be *underlined* or in *italic* type. Titles of articles, dissertations, poems, etc., should be given in quotation marks. You may use a short form after the first reference.

Not to be considered titles are: categories of subject matter (e.g., "their rendering of the Annunciation" but "Rogier's *Annunciations* on panel." In the latter, Annunciation is a title, although of several paintings); names acquired by tradition ("the Friedsam *Annunciation;* Friedsam is not the name of the work); names of persons represented in works of art ("Saint John twists on his column" but "the *Saint John* is exhibited in the Uffizi" since the latter *John* is the name of the work); names of buildings and things (Unicorn Tapestries, Lindisfarne Gospels, Hour of Catherine of Cleves, Isenheim Altarpiece).

In general, exclude "The" from treatment as part of a title.

For footnote references to titles of books, use an initial *The* (*The Sistine Chapel*), but delete *The* for titles of journals (*Art Bulletin*).

2. Numerals

Spell out numbers beginning sentences, including dates: e.g., "Seventeen seventy-six lives in American history."

Spell out numbers under one hundred, except in a statistical or similar discussion with numerous such numbers. Numbers above and below one hundred that pertain to the same category should be treated alike throughout a paragraph: "the archives have 140 letters, 42 invoices, . . ."

Use Arabic numbers in footnotes, in page numbers, in dates, and in measurements generally: foot–inch figures, percents, fractions, and ratios (i.e., 5:6). In measurements, type a space after the number: i.e., 45 in., 45 cm (no period after "cm").

Spell out centuries in articles ("seventeenth century"), but use arabic numerals ("17th century") in notes, book reviews, and captions.

For life dates, repeat the first two digits (1432–1480), but use only the last two digits in all other instances (1923–25), except: 1900–1905; 1805–6; 1793–1802.

Inclusive page numbers: pp. 21–28, 345–46, 105–9, 200–205, 1880–90, 12,345–47 (see *Chicago Manual of Style* 8.69).

Use Roman numerals for series and volume numbers, and Arabic for part, section, and chapter numbers, even when these are given in Roman numerals in the cited publication: "chapter 3"; "volume III of series II"

3. Capitalization

Capitalize names of works of art, including buildings (Cathedral of St. John the Divine, Albani Tomb, Baptistery of Pisa, Sforza Monument) and analogous terms (Lehman Collection), but not generic words in association with titles (tomb of Cardinal Albani, collection of Robert Lehman, church of Notre-Dame, Isenheim Altarpiece, but Lotto's Louvre altarpiece).

Lowercase references to parts of a book: appendix/app.; bibliography/bibliog.; book/bk.; cat. no.; fascicle/fasc.; figure/fig.; folio/fol.; introduction/intro.; line; new series/n. s.; note/n.; page/p.; plate/pl.; preface; series/ser.; signature; volume/vol. (Capitalize Fig. in reference to illustrations in your article.)

In titles of publications in English, cap. first word, last word, all nouns, pronouns, adjectives, verbs, adverbs, and subordinate conjunctions. Lowercase articles, coordinate conjunctions, and prepositions, and the *to* in infinitives.

The capitalization of foreign titles in general follows the capitalization of the language as it is used. (See the chapter on foreign languages in type in the *Chicago Manual of Style*.)

For capitalization of particles, follow the usage of the named individual or tradition: Tolnay, Schlosser, but de la Tour, d' Hulst, de Stael, von Blanckenhagen, Der Nersessian, Van Buren, van Gogh, van der Weyden (in general, lowercase the particle in European names).

In general, sharply delimited period titles are capitalized, whereas large periods and terms applicable to several periods are not:

Archaic
Baroque
Early and High Renaissance
Early Christian
Gothic
Greek Classicism of the
    fifth century (otherwise,
    classicism)
Imperial
Impressionism
Islamic
Mannerist

antique
antiquity
classicism (see above)
medieval
modern

Neoclassicism for the late
    eighteenth-century
    movement (otherwise,
    neoclassicism)
Post-Impressionism
Pre-Columbian
Rococo
Roman
Romanesque
Romantic period
Xth Dynasty
Middle Ages

neoclassicism (see above)
postmodern
prehistoric
quattrocento

Capitalize places with distinct and titled identities: Western Europe, the West, the Continent, Northern Italy, but western France.
Capitalize theological terms:

Apostles
Archangel Gabriel
Baptism
Benedictional
Child
Christ Child
Church Fathers
Crucifixion
Eucharist/Eucharistic
Evangelists

Incarnation
Infant
Judgment Day
Judgment of Solomon
Man of Sorrows
Mass
Massacre of the Innocents
Mother
Nativity
Original Sin

God the Father
Gospel Book
Heaven
Holy Communion
Immaculate Conception
Virtues and Vices (cap. each of them, e.g., Envy/*Invidia*)

Passion Play
Pontifical
Prophets and Sibyls
Scripture
Three Marys

In general, capitalize formally named theological terms, lowercase those generically referred to: archangels, birth and death of Christ, breviary, canon tables, communion, disciples, his birth (no capitalized pronominal adjectives), prayer book, sacrament.

4. Spelling and Hyphenation

a historical fact, not an aesthetic, archaeology
analyze, emphasize, but exercise
appendixes, indexes, but codices
avant-garde (noun and adj.)
bas-relief
black-figure (only as adj.)
catalogue, catalogue raisonné
draftsman
focusing, modeling, labeling, traveling
freestanding
frescoes, but halos, manifestos, torsos
ground plan
inquiry, insure

medium, media
mid-fourteenth century
millennia
molding
practice
preeminent, reevaluation, but re-creation
self-portrait
sketchbook
still-life (adj.), still life (noun)
terra-cotta (adj.), terra cotta (noun)
wall painting
watercolor
well-known (before noun); well known

Use English form for foreign place names; e.g., Florence not Firenze, Munich not München.

For possessives, add an apostrophe and an *s* for all names, including those ending in *s* or a sibilant (Keats's, Degas's, Eakins's, Marx's), except for Jesus', Moses', and Greek names (Xerxes'). See *Chicago Manual of Style*.

Insert all accents given in the foreign language, including those on capital letters. However, in French and Italian, accents are not used on capital letters.

5. *Abbreviations*

In general, use standard abbreviations in footnotes, but abbreviate sparingly in main text.

Cite books of the Bible by short title, usually one word. See *Chicago Manual of Style*.

Do not abbreviate journal titles: *Journal of the Warburg and Courtauld Institutes*, not *JWCI*.

Manuscript locations: Bibl. Nat. gr. and number; Bibl. Nat. lat. and number; Brit. Mus., Brit. Lib., Vat. lat.

Saint: Use Saint and the standard form of the name in English in referring to saints. For places, churches, etc., use the local form, abbreviating where possible (St. Louis, Mo.: St.-Denis: Ste.-Chapelle; S. Apollinare, S. Lucia, SS. Annunziata). Dates: January 6, 1980, in text: Jan. 6, 1980, and Jan. 1980, in notes.

State names: spell in full in text or when standing alone. Otherwise, use standard state abbreviations rather than the two-letter zip code form (e.g., Calif., Conn., Del., D.C., Ill., Mass., Mich., Mo., N.J., N.Y., Pa.).

6. *Italics*

Use roman type for scholarly Latin words and abbreviations: ca., cf., e.g., etc., but retain italics for *sic*, which should be in brackets, i.e., [*sic*].

Italicize words and phrases in a foreign language that are likely to be unfamiliar to readers: for example, *cire perdue: modello* (pl. *modelli*); *ricordo*. A full sentence in a foreign language should be set in roman type. Familiar words and phrases should be in roman type: for example, a priori; cause célèbre; élan; façade; in situ; mea culpa; oeuvre; papier-mâché; pentimento (pl. pentimenti); plein air; repertoire; trompe l'oeil.

7. *Quotations*

Quotations must be absolutely accurate and carefully transcribed. An ellipsis (three spaced dots) indicates words dropped within a sentence. A period and three spaced dots indicates a deletion between sentences.

Unless governed by fair use, authors must obtain permission to quote published material. In general, one should never quote more than a paragraph or two from a text or more than a few stanzas from a poem without permission. It is the responsibility of the author to obtain permission.

Extracts of more than 50 words should be typed without opening and closing quotation marks, *double spaced* in block form, i.e., indented one inch from left margin. Shorter quotations should be run into the text.

Foreign language quotations of more than a line or two should be translated into English in the text, unless the significance of the quotation will be lost. The original text may be included in a footnote only if it is unpublished, difficult to access, or of philological relevance to the article.

"Emphasis added" indicates your addition to quoted matter.

Brackets in quoted material indicate author's interpolation; in inscriptions they indicate letters lost through damage. Parentheses indicate letters omitted as the result of abbreviation in inscriptions. Elsewhere parentheses indicate the interpolation of the source, not the author quoting it.

Use the serial comma, as in Tom, Dick, and Harry.

Omit honorifics except when thanking the person for help. In general, omit honorifics when citing debts to published or older sources; give honorifics when citing current unpublished ones (letters, oral communication). Dr. but Professor, M., Mme, Mlle, Sig., Sig.ra, Signorina, Dott., Dott.ssa., Rev., Msgr.

Text References, Footnotes, and Frequently Cited Sources

References to biblical passages should be made in the text (e.g., Matt. 4:14), but the first citation to the Bible should be a full reference in a footnote, giving the version used, e.g., Vulgate, King James, Douai. Also to be cited in the text are other classic works with standardized systems of subdivision traditionally established, e.g., *Odyssey* 9.266, Timaeus 484b. The one full reference, in a footnote, should note which translation, critical edition, or the like was used.

For classic works that ought to be cited by edition, but exist in numerous editions, give extra reference to the section of the work (bk. IV, novella 3), since readers may use a different edition from yours.

Footnote reference numbers in the text should be clearly designated by means of superior figures placed after punctuation. Footnote reference numbers in the notes themselves should be on the same line as the text, followed by a period and one space (e.g., 1. John Shearman, . . .).

All references to publications and the like should appear in full form only once, and otherwise in a short form. Do not use *op. cit.* References will differ slightly depending on whether the work is cited only a few times in the article, or more often. In the latter case, the one full reference should be in the Frequently Cited Sources, at the end of the article, and all other references should take the short form of the author's surname and the page (e.g., Smith, 423), or the shortest comprehensible form (e.g., ASR, Cam., iii. Anagni, busta 80, fol. 40r).

If there is more than one author of the same surname, give an initial; in the case of more than one publication by an author, give the date (Smith, 1969, 423). In the case of more than one work by one author of the same date, cite one as "a" and one as "b" (to be specified also in the Frequently Cited Sources); thus the maximum short reference would be B. Smith, 1969a, 423.

If a reference is followed immediately by another to the same work, ibid. replaces "Smith" or any longer identification.

If a work is cited only a few times, it should not appear in the list of Frequently Cited Sources; the full reference should be in the first footnote reference to the work. Subsequent references should then take the short form, referring to the first reference: Smith (as in n. 15), 243.

In footnotes, the form of the full reference is as follows (the form for Frequently Cited Sources is the same, except that the author's surname is given first; i.e., Swindler, M. H.):

Formerly, *Art Bulletin* style required only initials for authors' given names. Current practice is to give names in full (John Rupert Martin rather than J. R. Martin). However, manuscripts prepared according to the old style are acceptable. Aim for consistency.

Asian names: The traditional format for Chinese and Japanese names is with family name first followed by the given name. Unless the name is Westernized, as it often is by authors writing in English, it should be kept in the traditional order. (Traditional order: Tsou Tang; Tajima Yumiko. Westernized: Tang Tsou; Yumiko Tajima.)

Book

1. Elizabeth Cropper, *Pietro Testa, 1612-1650,* exh. cat., Fogg Art Museum, Cambridge, Mass., 1988, 246. (For England, use Cambridge only.)

2. Annibale Caro, *Lettere familiari*, II, ed. A. Greco, Milan, 1957, 401-5.

3. William M. Smith, *Medieval Painting*, 2nd ed., Paris, 1925, 195-96.

4. Henry-Russell Hitchcock, *Architecture: Nineteenth and Twentieth Centuries* (1957), Pelican History of Art, rev. ed., Baltimore, 1966, 21-42.

5. *Annual of the American Academy in Rome*, III, Rome, 1922, passim. ("Passim" should be used sparingly; better to give inclusive page numbers.)

Article in periodical
1. Wilibald Sauerländer, "Die kunstgeschichtliche Stellung Westportale von Notre-Dame in Paris," *Marburger Jahrbuch für Kunstwissenschaft*, XVII, 1959, 1-55.

2. Jan Jennings, "Leila Ross Wilburn, Plan-Book Architect," *Woman's Art Journal*, Spring-Summer 1989, 15.

3. Antonio Natali, "Altro da Pontormo e Bronzino?" *Antichità Viva*, nos. 2-3, 1989, 136-37.

Distinguish between vol. and no.: *Art Bulletin*, LII, no. 3 [note: no *The* in titles of journals].

Unpublished material
1. Nancy Elizabeth Locke, "Manet and the Family Romance," Ph.D. diss., Harvard University, 1993. 41; or, M.A. thesis, New York University, Institute of Fine, Arts, 1961, 41.

2. Klaus Bauch, *Bildnisse des Jan van Eyck*, University of Munich, 1970 (because German dissertations are published).

Archival material
1. Rouen, Bibli. Mun. MS fr. 938 (comma only between city and institution).

2. London, Brit. Mus. Add. MS 28134 (not B.M. or BM); or invert sequence at author's discretion: Archivio di Stato, Rome (hereafter ASR), Camerale III. Anagni, busta 80,

Computa Depositaria Munimimis Ananiae, 3 (cap. as given by author; no italics).

For certain standard reference tools, the first reference may use a short title, notably Thieme-Becker, *Pat. lat., Pat. gr.,* Vasari-Milanesi, *Webster's Third, Webster's International.* In arranging the list of Frequently Cited Sources, archival material, if of any length, should be cited first, showing abbreviations as used in the footnotes for names of archives and the like, e.g.:

ASR Archivio di Stato, Rome

Any other abbreviations, unless numerous, should be alphabetized within the main list. References should appear alphabetically by author. For alphabetization, see *Webster's Biographical Dictionary* or *Who's Who in American Art.* Some examples:

Amstel, J. van
Annual of the British Academy in Athens, LVII, London, 1972
Blanckenhagen, P. von
Busiri Vici, A.
Der Nersessian, S.
Metropolitan Museum of Art, *The Year 1200,* New York, 1975
Van Buren, A. H.

When there are many contributions by the same author in the list of Frequently Cited Sources, they should be organized by date, placed after the author's name, with earliest date cited first:

Meiss, Millard, 1956, "Jan van Eyck," *Venezia e l'Europa,* Venice, 58-69.
————, 1967, "Sleep in Venice," *Stil und Uberlieferung,* Berlin, 100-120.
————, 1970, "The Friedsam Annunciation Again," *Art Bulletin,* LII, 368-72.

Note that inclusive page numbers of articles must be provided in the list of Frequently Cited Sources.

Captions

Captions should be numbered consecutively. Since authors are responsible for obtaining reproduction permissions, captions will reflect the institution's requirements and those of the photographer. Provide the shortest possible form: by the gracious permission of Her Majesty the Queen [equals] by permission H. M. the Queen. Sample captions (figure numbers are set in boldface, without periods):

1 Castiglione, *Crucifixion*. Genoa, Palazzo Bianco (photo: Frick Art Reference Library)

2 Pietro Bernini, Bust of Scipio (or *Scipio*). Rome, S. Giovanni dei Fiorentini (photo: Davis Lees, Rome)

3 Attributed to Cherubino Alberti, engraving after Florence Cathedral *Pietà*, ca, 1572. Vienna, Albertina (courtesy J. Held)

4 Parthenon, east frieze (detail) (from Smith, *The Parthenon,* pl. 2) (if Smith is not in notes or list of Frequently Cited Sources, give full citation: A. Smith, *The Parthenon,* New York, 1975, pl. 2)

5 *Tree of Vices, Le Verger de Soulas,* northern French, ca. 1290. Paris, Bibl. Nat. MS fr. 9220, fol. 10r (photo: Bibl. Nat.)

6 Roman sarcophagus, *Death of Meleager* (detail). Paris, Louvre (photo: Alinari)

For many illustrations from the same source and where space is tight and images small, use the full citation for the first illustration, then shorten. For example:

1 Little Canterbury Psalter, *Nativity.* Paris, Bibl. Nat. MS lat. 770, fol. 20r (photo: Bibl. Nat.)

2 Little Canterbury Psalter, *Ascension,* fol. 36v.

CORRECTIONS IN THE FINAL COPY

Your extensive revisions should have been made in your drafts, but minor last-minute revisions may be made on the finished copy. In proofreading you may catch some typographical errors, and you may notice some minor weaknesses. It's not a bad idea to read the paper aloud to someone, or to yourself. Your tongue will trip over phrases in which a word is omitted, or sentences that are poorly punctuated. Some people who have difficulty spotting errors report that they find they are helped when they glide a pencil at a moderate speed letter by letter over the page. Without a cursor, they say, their minds and eyes read in patterns and they miss typos.

Let's say that in the final copy you notice an error in agreement between subject and verb: "The weaknesses in the draftsmanship is evident." The subject is "weaknesses" (not "draftsmanship") and so the verb should be "are," not "is." You need not retype the page or even erase. You can make corrections with the following proofreader's symbols.

Changes in wording may be made by crossing through words and rewriting just above them, either on the typewriter or by hand in ink or colored pencil:

The weaknesses in the draftsmanship ~~is~~ *are* evident.

Additions should be made above the line, with a caret ($_\wedge$) below the line at the appropriate place:

The weaknesses in the draftsmanship $_\wedge$ *are* evident.

Transpositions of letters may be made thus:

The weaknesses in the draftsmanship are evident.

Deletions are indicated by a horizontal line through the word or words to be deleted. Delete a single letter by drawing a vertical or diagonal line through it.

The weaknesses in ~~in~~ the draftsmanship are evident.

Separation of words accidentally run together is indicated by a vertical line, **closure** by a curved line connecting the things to be closed up:

The weaknesses|in the d̂raftsmanship are evident.

Paragraphing may be indicated by the symbol ¶ before the word that is to begin the new paragraph.

The weaknesses are evident.¶For instance, the draftsmanship is hesitant, and the use of color is. . . . .

10

Essay Examinations

WHAT EXAMINATIONS ARE

The first two chapters of this book assume that writing an essay requires knowledge of the subject as well as skill with language. Here a few pages will be devoted to discussing the nature of examinations; perhaps one can write better essay answers when one knows what examinations are.

A well-constructed examination not only measures learning and thinking but also stimulates them. Even so humble an examination as a short-answer quiz is a sort of push designed to move students forward by coercing them to do the assigned looking or reading. Of course, internal motivation is far superior to external, but even such crude external motivation as a quiz can have a beneficial effect. Students know this; indeed, they often seek external compulsion, choosing a particular course "because I want to know something about . . . and I know that I won't do the work on my own." (Instructors often teach a new course for the same reason; we want to become knowledgeable about, say, the Symbolists, and we know that despite our lofty intentions we may not seriously confront the subject unless we are under the pressure of facing a class.)

In short, however ignoble it sounds, examinations force students to acquire knowledge and then to convert knowledge into thinking. Sometimes it is not until preparing for the final examination that students—returning to museums, studying photographs of works of art, rereading the chief texts and classroom notes—see what the course was really about; until this late stage the trees obscure the forest, but now, reviewing and sorting things out—*thinking* about the facts, the data, and the ideas of others—they see a pattern emerge. The experience of reviewing and then of writing an examination, though fretful, can be highly exciting, as connections are made and ideas take on life. Such discoveries about the whole subject matter of a course can almost never be made by writing critical essays on topics of one's own construction, for such topics rarely require a view of the whole. Further, we are more likely to make imaginative leaps when trying to answer questions that other people pose to us than when trying to answer questions we pose to ourselves. (Again, every teacher knows that students in the classroom ask questions that stimulate the teacher to see things and to think thoughts that would otherwise have been neglected.) And although a teacher's question may cause anxiety, when students confront and respond to it on an examination they often make yet another discovery—a self-discovery, a sudden and satisfying awareness of powers they didn't know they had.

WRITING ESSAY ANSWERS

Let's assume that before the examination you have read the assigned material, marked the margins of your books (but not of the library's books), made summaries of the longer readings and of the classroom comments, visited the museums, reviewed all this material, and had a decent night's sleep. Now you are facing the examination sheet.

Here are some obvious but important practical suggestions:

1. Take a moment to **jot down, as a sort of outline or source of further inspiration, a few ideas that strike you** after you have thought a little about the question. You may at the outset realize there are, say, three points you want to make, and unless you jot them down— three key words will do—you may spend all the allotted time on one point.

2. **Answer the question:** If you are asked to compare two pictures, compare them; don't write two paragraphs on the lives of each painter. Take seriously such words as *analyze, compare, summarize,* and especially *evaluate.*

3. **You can often get a good start if you turn the question into an affirmation**—for example, by turning "In what ways is the painting of Manet influenced by Goya?" into "Manet's painting is influenced by Goya in at least . . . ways."

4. **Don't waste time summarizing** at length what you have read unless asked to do so—but, of course, you may have to give a brief summary in order to support a point. The instructor wants to see that you can *use* your reading, not merely that you have done the reading.

5. **Budget your time.** Do not spend more time on a question than the allotted time.

"I think you know everybody."

6. **Be concrete.** Illustrate your argument with facts—names of painters or sculptors or architects, titles of works, dates, and brief but concrete descriptions.

7. **Leave space for last-minute additions.** Either skip a page between essays, or write only on the right-hand pages so that on rereading you can add material at the appropriate place on the left-hand pages.

Beyond these general suggestions, it is best to talk about essay examinations by looking at the most common sorts of questions:

- a work to analyze
- a historical question (e.g., "Trace the influence of Japanese art on two European painters of the later nineteenth century"; "Trace the development of Picasso's representations of the Minotaur")

- a critical quotation to be evaluated
- a comparison (e.g., "Compare the representation of space in the late works of van Gogh and Gauguin")

A few remarks on each of these types may be helpful.

1. **On analysis,** see Chapter 2. As a short rule, look carefully at subject matter, line, color (if any), composition, and medium.

2. A good essay on a **historical question,** like a good lawyer's argument, will offer a nice combination of argument and evidence; that is, the thesis will be supported by concrete details (names of painters and paintings, dates, possibly even brief quotations). A discussion cannot be convincing if it does not specify certain works as representative—or atypical—of certain years. Lawyerlike, you must demonstrate (not merely assert) your point.

3. If you are asked to evaluate a **critical quotation,** read the quotation carefully and in your answer take account of *all* of the quotation. If, for example, the critic has said, "Goya in his etchings usually . . . but in his paintings rarely . . . ," you will have to write about etchings and paintings (unless, of course, the instructions on the examination ask you to take only as much of the quotation as you wish). Watch especially for such words as *usually, for the most part, never;* that is, although the passage may on the whole approach the truth, you may feel that some important qualifications are needed. This is not being picky; true evaluation calls for making subtle distinctions, yielding assent only so far and no further. Another example of a quotation to evaluate: "Picasso's *Les Demoiselles d'Avignon* [illustrated in this book on page 00] draws on several traditions, and the result is stylistic incoherence." A good answer not only will specify the traditions or sources (e.g., Cézanne's bathers, Rubens's *The Judgment of Paris,* Renaissance and Hellenistic nudes, pre-Christian Iberian sculptures, Egyptian painting, African art), calling attention to the passages in the painting where each is apparent, but also will evaluate the judgment that the work is incoherent. It might argue, for example, that the lack of traditional stylistic unity is entirely consistent with the newness of the treatment of figures and with the violent eroticism of the subject.

4. **On comparisons,** see Chapter 3. Because lucid comparisons are especially difficult to write, be sure to take a few moments to jot down a sort of outline so that you know where you will be going. You can often make a good start by beginning with the similarities of the two objects. As you jot these down, you will probably find that your perception of the *differences* will begin to be heightened.

Hans Namuth, *Jackson Pollock Painting*, 1950. (© Estate of Hans Namuth; courtesy of the Pollock-Krasner House and Study Center, East Hampton, NY)

In organizing a comparison of two pictures by van Gogh and two by Gauguin, you might devote the first and third paragraphs to van Gogh, the second and fourth to Gauguin. Or you might first treat one painter's two pictures and then turn to the second painter. Your essay will not break into two parts if you announce at the outset that you will treat one artist first, then the other, and if you remind your reader during your treatment of the first artist that certain points will be picked up when you get to the second, and, finally, if during your treatment of the second artist you occasionally glance back to the first.

LAST WORDS

Chance favors the prepared mind.

—Louis Pasteur

Credits

Arnheim, Rudolph. Quotation and drawing from *Art and Visual Perception* (1974). Reprinted by permission of the University of California Press.

Bedell, Rebecca, "Mrs. Mann and Mrs. Goldthwait" by Rebecca Bedell. Copyright © 1981 by Rebecca Bedell. Used by permission of the author.

College Art Association. *Art Bulletin Style Guide*, pp. 3–11. Reprinted by permission of the College Art Association.

Elsen, Albert E. Excerpt from *Purposes of Art*, Third Edition. Copyright © 1972 by Holt, Rinehart and Winston, Inc., by permission of the publisher.

Herbert, Robert. "Millet's *The Gleaners*," reproduced from the exhibition catalogue *Man and His World: International Fine Arts Exhibition.* © The National Gallery of Canada for the Corporation of the National Museums of Canada. Reprinted by permission of the National Museums of Canada.

Johnson, Eugene, J. Quotation in *International Handbook of Contemporary Developments in Architecture*, edited by Warren Sanderson. Copyright © 1981 by Warren Sanderson. Reprinted by permission of Greenwood Press, Westport, CT.

Kinney, Leila W. "The Internet and Research" was written for this book and is used by permission of the author.

McCauley, Elizabeth Anne. "Photography" was written for this book and is used by permission of the author.

Pemberton, John III. Excerpt from "The Carvers of the Northeast" in *Yoruba: Nine Centuries of African Art and Thought*, edited by Allen Wardwell (New York: Center for African Art, 1989), p. 206, reprinted by permission of John Pemberton III.

Saslow, James, M. "Gay and Lesbian Art Criticism" was written for this book and is used by permission of the author.

Index

Symbols Commonly Used in Annotating Papers

All instructors have their own techniques for commenting on essays, but many make substantial use of the following symbols. When instructors use a symbol, they assume that the student will carefully read the marked passage and will see the error or will check the appropriate reference.

awk(k)	awkward
cap	use a capital letter
cf	comma fault
choppy	too many short sentences; subordinate (see pp. 135–137)
diction	inappropriate word (see p. 129)
emph	emphasis is obscured
id	unidiomatic expression
ital	underline to indicate italics (see pp. 234, 258)
k	awkward
l	logic; this does not follow
lc	use lowercase, not capitals
mm	misplaced modifier
¶	new paragraph
pass	weak use of the passive voice (see pp. 133–134)
ref	pronoun reference is vague or misleading
rep	awkward repetition (see pp. 130–31)
sp	misspelling
sub	subordinate (see pp. 135–37)
trans	weak transition (see pp. 141–42)
wordy	use concise language (see pp. 132–34)
ww	wrong word (see p. 126)
x	this is wrong
?	really? are you sure? I doubt it (or I can't read this)